Make:
Musical Inventions

Make: Musical Inventions

DIY Instruments to Toot, Tap, Crank, Strum, Pluck, and Switch On

By Kathy Ceceri

Musical Inventions

DIY Instruments to Toot, Tap, Crank, Strum, Pluck, and Switch On

By Kathy Ceceri

Printed in Canada

Published by
Maker Media, Inc.,
1700 Montgomery Street, Suite 240
San Francisco, California 94111

Maker Media books may be purchased for educational, business, or sales promotional use. Online editions are also available for most titles (*safaribooksonline.com*). For more information, contact our corporate/institutional sales department: 800-998-9938 or *corporate@oreilly.com*.

Publisher: Roger Stewart
Editor: Patrick Di Justo
Copy Editor: Rebecca Rider, Happenstance Type-O-Rama
Proofreader: Elizabeth Welch, Happenstance Type-O-Rama
Interior Designer and Compositor: Maureen Forys, Happenstance Type-O-Rama
Cover Designer: Maureen Forys, Happenstance Type-O-Rama
Illustrator: Richard Sheppard, Happenstance Type-O-Rama
Indexer: Valerie Perry, Happenstance Type-O-Rama

April 2017: First Edition
Revision History for the First Edition
2017-04-26 First Release

See *oreilly.com/catalog/errata.csp?isbn=9781680452334* for release details.

978-1-6804-5233-4

Safari® Books Online

Safari Books Online is an on-demand digital library that delivers expert content in both book and video form from the world's leading authors in technology and business. Technology professionals, software developers, web designers, and business and creative professionals use Safari Books Online as their primary resource for research, problem solving, learning, and certification training. Safari Books Online offers a range of plans and pricing for enterprise, government, education, and individuals. Members have access to thousands of books, training videos, and prepublication manuscripts in one fully searchable database from publishers like O'Reilly Media, Prentice Hall Professional, Addison-Wesley Professional, Microsoft Press, Sams, Que, Peachpit Press, Focal Press, Cisco Press, John Wiley & Sons, Syngress, Morgan Kaufmann, IBM Redbooks, Packt, Adobe Press, FT Press, Apress, Manning, New Riders, McGraw-Hill, Jones & Bartlett, Course Technology, and hundreds more. For more information about Safari Books Online, please visit us online.

How to Contact Us

Please address comments and questions concerning this book to the publisher:

Maker Media, Inc.
1700 Montgomery, Suite 240
San Francisco, CA 94111
877-306-6253 (in the United States or Canada)
707-639-1355 (international or local)

Maker Media unites, inspires, informs, and entertains a growing community of resourceful people who undertake amazing projects in their backyards, basements, and garages. Maker Media celebrates your right to tweak, hack, and bend any Technology to your will. The Maker Media audience continues to be a growing culture and community that believes in bettering ourselves, our environment, our educational system—our entire world. This is much more than an audience, it's a worldwide movement that Maker Media is leading. We call it the Maker Movement.

For more information about Maker Media, visit us online:

Make: and Makezine.com: *makezine.com*
Maker Faire: *makerfaire.com*
Maker Shed: *makershed.com*

To comment or ask technical questions about this book, send email to *bookquestions@oreilly.com*.

The following are attributions and photography credits for images included in this book:

Ben Franklin's Glass Armonica Dean Shostaks Williamsburg VA 3384 via the Creative Commons (*https://creativecommons.org/licenses/by/2.0/*) | Figure 3-10: Martin Molin playing his Marble Machine, Photo credit: Samuel Westergren, courtesy of Wintergatan | Figure 3-34: Experimental Music Box courtesy of Koka Nikoladze (*nikoladze.eu*) | Figure 3-67: A musical invention by Pierre Bastien. Photo Credit: Julie Vermeeren | Figure 4-8: Photo credit: Jade Soto, used by permission | Leon Theremin: Photo credit: Copyright free, via Wikimedia | Figure 4-28: Photo credit: Hannah Perner-Wilson/mi.mu gloves | Figure 4-43: Photo credit: Reed Ghazala | Thomas Edison: Copyright free, via the Library of Congress | Grand Wizzard Theodore DJing at Experience Music Project in Seattle, WA. Photo by Flickr user Flintmi via the Creative Commons (*https://creativecommons.org/licenses/by/3.0/deed.en*) | Pierre Curie: Photo by Dujardin c 1906 entitled, Traité de radioactivité. edition, Paris: Gauthier, 1910 | Guglielmo Marconi: Copyright free, via the Library of Congress | Nikola Tesla: Copyright free, via Wikimedia

Contents

Acknowledgments

I would like to thank the following people for their help in the writing of *Musical Inventions*:

- Marilynn Beuhler and Mac Petrequin, for their advice and feedback
- Rebecca Angel Maxwell and Ayah and Thea Goldman for testing and modeling my projects
- Deborah Segel (*violinsdirect.com*) for demonstrating the use of the singing bowls
- Keith Handy (*soundcloud.com/keithhandy*, *vimeo.com/keithhandy*)
- Dave Barnes (*youtube.com/user/FREAKENSPEAK*)
- Balam Soto (*balam.io*)
- Hannah Perner-Wilson (*mimugloves.com*) for photos and information about Imogen Heap's mi.mu gloves
- Ayah Bdeir and Liza Stark (*littlebits.cc*) for help with the littleBits Synth glove
- Chuck Porter and Connie LaPorta for help and advice with the crystal radio

My thanks also go to the people at Maker Media and Happenstance Type-O-Rama who helped get the book from my work table to your hands:

- Patrick DiJusto
- Gretchen Giles
- Roger Stewart
- Maureen Forys
- Rebecca Rider

Preface: Musical Inventors

There is a long tradition of turning recycled items into homemade instruments. But creating new instruments is much more than an art form. It take science, technology, engineering, and math to build an instrument that sounds good, looks good, works well, and is enjoyable to play. And it takes a couple other things—a love of music, and a sense of fun.

FIGURE P-1: Bluegrass musician, math teacher, and musical inventor Mac Petrequin makes banjos out of cookie tins in his basement workshop.

I wrote *Musical Inventions* because I'm a Maker who loves to explore ways to turn everyday stuff into fun and useful objects. I get excited when I start to understand why things like musical instruments behave the way they do. Of course, I also like to play music: I studied the violin in school and later learned how to play the piano and the mandolin. But like most people, I'm not an expert at playing or making musical instruments, so the projects in this book are all aimed at beginners. All the projects are kid-friendly as well, although some will require adult help and supervision.

Even though the projects are simple, many produce playable instruments that are good enough to let you pick out a tune. Others create interesting sounds that you can build upon to design your own kind of musical invention. The last chapter will explain how technology like the phonograph and the radio actually changed the kind of music people listened to and will show you how to make some simple versions of your own. All the projects will teach you about science and technology, as well as how new inventions make their way into popular culture.

As you read through *Musical Inventions*, keep an eye out for short profiles of other inspirational musical inventors from the past and present. One thing you will notice from their stories is that creating something new isn't always easy. As new media artist Balam Soto says in Chapter 4, the invention process often involves testing your design, making mistakes, and starting over. But if you stick with it, being a Maker—especially a musical Maker—can be very rewarding.

I hope you enjoy learning about how instruments work and how to create your own. I always love seeing what readers have created, so feel free to send me photos of your own musical inventions and I will share them with my audience. You can reach me through my website, *craftsforlearning.com*, where you will also find videos of project prototypes.

Happy Making!

—Kathy Ceceri

INTERVIEW WITH A MUSICAL MAKER

To help me write this book, I spoke to real musical inventors to find out what they do. One was musician Keith Handy. Keith teaches circuit bending techniques to kids and other beginners. *Circuit bending* (which you will learn about in Chapter 4) involves taking apart an old electronic device, such as a talking toy, and turning it into a one-of-a-kind music synthesizer by changing around the wiring inside to alter the kinds of sounds it can make. His story is shared by a lot of modern-day musical inventors.

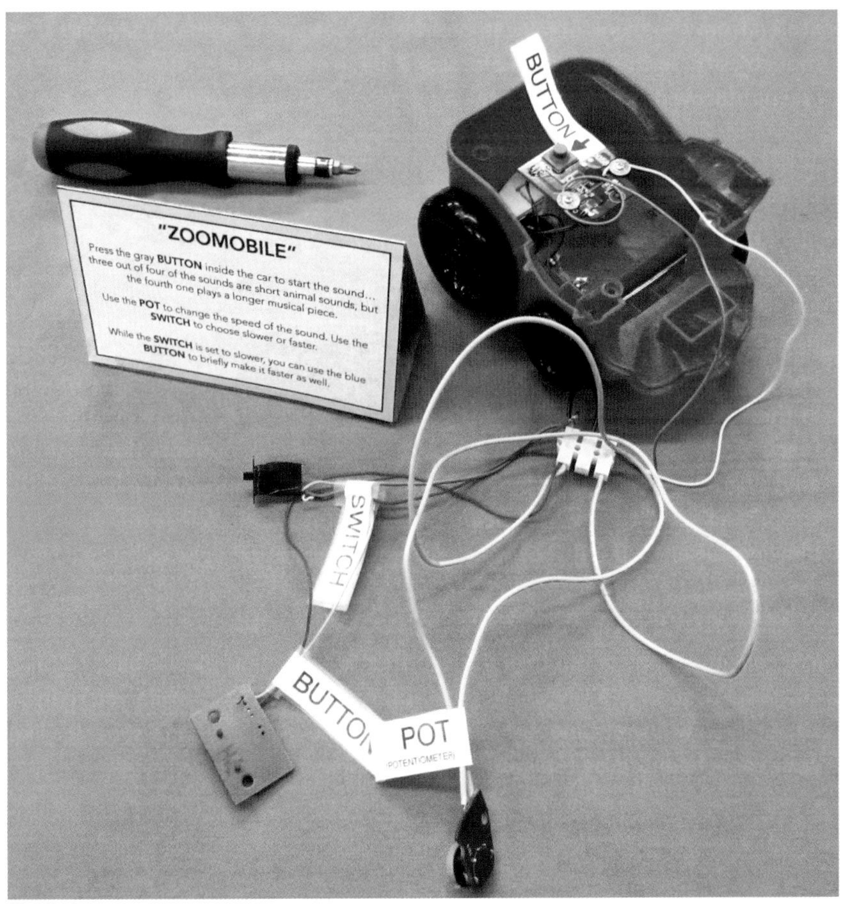

FIGURE P-2: A talking toy car turned into a demonstration of circuit bending by musician Keith Handy.

How did you get started inventing instruments?

Pretty much every electronic thing I've ever owned—toys, tape recorders, synthesizers—has been dismantled, if I could get into it with a Phillips-head screwdriver. This almost seems to be to be a primal impulse. You have a thing and you just want to see what's inside it. Lots and lots of trying to alter and modify devices, mostly failing, and occasional succeeding. I've never been able to put something back together properly. When I was about 17, I took apart two of my keyboards so I could wire a full-size keyboard, note by note, to the other instrument that originally had a smaller-scale keyboard. The whole thing was mounted to a giant wooden board and that's what I played on stage until it was stolen from the trunk of my car. Somebody out there probably still has that monstrosity. To this day, if an electronic device has passed its era of usability, I will not throw it away without taking it apart first. It's a ritual.

What do you like (and dislike) about the trial-and-error process involved in circuit bending a device?

I like that it's an inroad to learning electronics, not starting from the basics, but starting from play. Play leads to questions, and questions lead to learning. It's a lot more fun for the beginner than reading about topics like "What does *current* mean?" I also just like seeing the inside of things. You don't immediately know what everything does, but you know everything was put in there deliberately, and has a reason for being there, and to me that's fascinating.I don't like finding tiny surface-mount components that I can't do anything with. And it's no fun when you fry a component, maybe hear a little "pop," and smell the tell-tale scent of electronic death. A lesser failure is blowing out an LED. The toy will keep working but that light will stay dark.

What funny stories can you share?

We collect dozens of old toys from thrift stores in giant plastic bags to take to classes. Frequently a toy will bump another toy in the bag while you're in transit, and you will spontaneously hear a silly voice or a children's song. It's like your car is haunted by the ghost of Walt Disney.

The Musical Inventions Supply Closet

Many of the projects in this book can be made using only common household items and crafts materials. I assume you already have things like scissors and paper or a hammer and nails, or that you know where to find them easily.

The items below include tools and materials that you may not have on hand, and the lists give you some suggestions for what kind of materials to get and where to find them.

Don't worry, you won't need all these items for every project! In fact, in some cases, you may be able to make do with what you have on hand. For instance, you can cut wire, and even strip off the plastic insulation from a wire, with a regular pair of utility scissors that you use for cutting different materials. It won't be as quick and clean as it would be with a dedicated wire cutter and wire stripper, and you may have to try it a few times to do it right, but in a pinch, using scissors will work almost as well. Also, some items—such as a metal mixing bowl or a portable device (smart phone, mp3 player) for listening to music—are only needed temporarily, so you can use what you already own.

And for any of the materials you need for this book, it's OK to use items you find at thrift shops and garage sales. Ask friends and relatives if they have any of these materials tucked away in their garage or basement that they don't need any more. Or try eBay, Craigslist, or local freeshare groups. Be creative, and you won't have to spend a lot of money to become a Musical Inventor!

Note: Check my website, *craftsforlearning.com*, for links to kits and parts bundles for specific projects in this book!

Here are some of the things you'll need:

Thrift Shop Materials

Although some of these are only borrowed temporarily for the project, such as bowls, it's always safer to look for used items that you can build into projects or that you can take apart without worrying about destroying them.

- Stemmed wine glasses
- Old wrenches, various sizes
- Old bowls, various sizes
- Old keys
- Old LP records (throwaways, not cherished classics—they may become damaged)
- Wooden cigar box
- Small tabletop electric fan
- Battery-powered AM (or AM/FM) radios (at least 3)
- Remote control (RC) car
- Plug-in (not wireless) mp3 player amplifier

Dollar Store/Discount Department Store/Supermarket Materials

Cheap versions of these items are fine for the projects in this book.

- Nylon fishing line, different weights
- Plastic straws, assorted widths
- Cucumbers, carrots, and assorted vegetables
- Apple corer (from the cooking utensil aisle)
- Mini wind chimes
- Large bobby pins (in the hair care aisle)
- Bamboo skewers
- Mini bells
- Glass marbles
- Wooden rolling pin

- Musical greeting cards (look for American Greetings)
- Sewing needles (or pins)
- Large belt rivets (from the sewing section)
- Earbuds (several pairs, as cheap as possible since they may be cut up and built into the project)

Portable device for listening to music, such as an mp3 player or FM or AM transistor radio (it will not be built in permanently, so you can use your smart phone or tablet)

- Plug-in cell phone/mp3 player speakers (not wireless)
- 9V (9 volt) battery

Recycling Bin Materials

Hold onto these items if they come through your recycling basket.

- Extra-long cardboard tubes from wrapping paper
- Cardboard tubes from paper towels
- Empty cookie tins
- Metal coffee cans or large metal or cardboard cans with metal bottoms
- Metal bottle caps

Art Store/Office Supply Store Materials

You probably have a few of these lurking in the bottom of your junk drawer.

- Pencils (with erasers)
- Wooden craft sticks of various sizes (coffee stirrers, popsicle sticks, tongue depressors, etc.)
- Wooden beads
- Pushpins
- Thumbtacks
- Index cards
- Soft lead pencils (look for art pencils marked 2B, 4B, or higher)

- Hot glue gun (low-temp is fine)
- Manila folders
- Clear wide packing tape
- Brass brads

Hardware Store Materials

In addition to the items I list here, you may need tools like hammers and screwdrivers in various sizes. An electric or manual drill is also handy. For nuts, bolts, washers, wood pieces, and so on, check project directions to see if you need exact sizes.

> **Safety Warning:** Children should only use tools, manual or electric, with adult supervision!

- Small metal pail
- Eyebolts and matching nuts
- Nails (for poking holes in cans and other uses)
- Assortment of washers (flat metal disc-like rings)
- Wooden screw-in extension handle for a paint roller
- Rubber cane tip
- Thin, soft nylon cord (enough to use in several projects)
- Sandpaper (medium and fine)
- Utility knife (box cutter)
- Scrap wood, especially small square boards and 1×2 strips (which are actually closer to $\frac{3}{4} \times 1\frac{1}{2}$ inch (19×38 mm)—pine is easy to work with, hardwood is sturdier; look for scraps in the lumber department or ask them to cut a piece to size for you)
- Assorted wood screws
- Corrugated tubing (such as drainage hose from the plumbing aisle), assorted widths
- Wooden paint stirrers (sometimes free in the paint department)

- Assorted small flexible springs (can also be salvaged from ball point pens and other household items)
- Aluminum foil tape (in the heating duct aisle)
- Wire cutters
- Wire strippers

Electronics Parts and Kits

Look for these items at your local electronics supply store or on Amazon and eBay, retail websites such *as www.sparkfun.com, www.adafruit.com*, and *www.jameco.com*, or science education retail websites like *miniscience.com*. Radio Shack (if there is still a store near you!) has been a good source for some of these items, but they tend to be more expensive.

- littleBits Synth kit and additional modules (*littlebits.cc*)
- Copper tape with conductive glue
- Makey Makey (*www.makeymakey.com*)
- Alligator clip wires (also known as test leads) *Note:* If you have a Makey Makey kit, it includes alligator clip wires that you can use for the other projects in this book.
- Magnet wire, 22, 26, or 30 gauge (Radio Shack sells a three-pack with these sizes)
- Crystal radio earphone
- Germanium diode (1N34A)
- Additional wire (several yards/meters) for an antenna and ground wire, if needed, for the radio project (hook-up wire is easy to work with, but almost any kind of strong thin wire will do)

Note: Keep an eye out for clearance sales on kids' electronics kits, especially crystal radio kits. Even if they are not high quality, they can be a cheap source of hard-to-find electronics parts to build the projects in this book.

Introduction: The Science of Sound and the Art of Inventing Instruments

Music is a fundamental part of our evolution; we probably sang before we spoke.

—JAY SCHULKIN, *Reflections on the Musical Mind*

FIGURE I-1: **You can make almost anything into an instrument. Some musical inventor turned this chemistry set into an electric ukulele.**

Nobody knows who invented music. But music may have been one of the earliest forms of human communication. That's because music can tell stories and create moods without using a single word. Music can make you laugh or cry, lull you to sleep, or send you marching off into battle. Music's ability to bring people together may have helped early humans form social groups, build cities, and forge civilizations. One thing is certain—from the start, music has been one of the most powerful forces in human history.

The first musicians didn't need anything besides their voice, hands, and feet to create their rhythms and melodies. When they began to make instruments, they turned to materials around them—sticks, bones, seashells, and animal horns. One of the oldest musical instruments ever found was a flute made from the tusk of a wooly mammoth, an extinct species of elephant that roamed Europe during the most recent ice age. When scientists discovered the flute in a cave in Germany in 2012, tests showed that it was more than 42,000 years old—thousands of years older than the earliest known cave paintings.

In this book, you'll be introduced to many kinds of instruments, from primitive flutes to the latest kinds of electronic synthesizers. You'll also find out how to create your own! When you're the inventor, you get to decide how the instrument looks and sounds. But to come up with any new instrument, it helps to understand how instruments really work. And for that, you have to understand the science of sound, a few music basics, and how they are related.

What Is Sound?

Sound is created by vibrations. An object *vibrates* when it shakes back and forth over and over. When you tap, pluck, rub, jiggle, or blow air across an object, you are transferring a little bit of your body's energy into it, and into the tiny particles called molecules that make up every form of matter. This added energy can make the molecules vibrate—and those molecules bump into nearby molecules and make them start vibrating too. If you've ever felt the floor thump when music is played on big bass speakers, you've felt sound energy traveling through solid material.

THE SCIENCE OF MATTER, ENERGY, AND ELECTRICITY FOR MUSICIANS

In order to understand the science of sound, it helps to understand what the universe is made of. Yes, really! Here's the short version: the part of the universe that we deal with every day consists of matter and energy. *Matter* is anything that takes up space. It is made up of atoms. An *atom* is the smallest piece of a substance that still behaves like that substance. Atoms can be joined together in a group, which can include different kinds of atoms.

The smallest group of atoms that shows its own kind of behavior is called a *molecule*. For example, hydrogen and oxygen are both gases. They float around in the air. But if you take two atoms of hydrogen and combine them with one atom of oxygen, you get a new substance—water! Although it is made up of two gases, at room temperature water is a liquid that flows but does not float (except in tiny droplets of water vapor). Its properties, the way it behaves, are different. A cup of water is made up of billions of water molecules—and each one is a tiny group of two hydrogens and one oxygen, H_2O.

Atoms are made up of smaller particles. It's the different combinations of particles that make each atom different. All atoms contain a *nucleus* (made up of several kinds of particles we won't go into here), surrounded by a cloud of much smaller particles called *electrons*. Each of these has a charge. The nucleus of an atom has a positive charge, and an electron has a negative charge. *Charge* is hard to define—all you need to know is that opposite charges attract each other, and charges that are the same push each other away (just like the positive and negative poles on a pair of magnets).

Electrons have the ability to move from one atom to another, carrying their negative charge with them. The flow of negative charge is *electricity*. To make electricity run your devices and do other kinds of work, you need to move it where you want it to go. A path designed to carry flowing electricity is called a *circuit*. To help the flow along, a circuit should be *closed*—in other words, it should be connected in a loop. When a circuit is closed, each atom can give up an electron in front of it and grab one from the atom behind it, allowing the electrons to flow around and around.

To shut off the electricity, all you have to do is open a switch so the electrons can't flow anymore. In some of the projects in this book, you will sometimes be asked to connect a circuit to a ground. That's just a way of closing the circuit. However, in the radio project in Chapter 5, you may literally ground your circuit, by way of a pipe that goes into the earth below your building. The earth absorbs the electrons being sent through the wires, creating one gigantic circuit!

Circuits are made of materials that give up their electrons easily. Materials that let electricity flow easily are called *conductive*. They include metals such as copper. Materials that don't carry electricity easily are said to have a high *resistance*. They are used to make resistors that slow down the flow of electrons. In Chapter 4, you build a resistor using pencil lead (graphite), which is a conductor, but not a good one. Materials that keep electricity from flowing completely are called *insulators*. You can use insulated wires and paper to prevent electricity from going where you don't want it to and causing a short circuit.

Energy is what makes matter move and change. There are different kinds of energy, and you can change one kind of energy into another. Energy can also be changed into matter and back again! You will find that electrons are sometimes described as little particles that move along, like the flow of electrical charge in a circuit. Other times, electrons behave like energy, and flow in waves that can travel through the vacuum of space. The reason that both are true is because energy and matter are related. (You may have heard of the formula for converting them: $E=mc^2$.)

It's this astounding fact that makes it possible to do all kinds of cool things, like changing electrical energy (from radio waves, for example) into mechanical energy (sound waves that make air molecules move back and forth). Keep reading to find out how!

Sound vibrations can also travel through the air. Air is also made up of molecules, and when they start to vibrate, they can carry the sound energy right to your ear. Inside your ear, your eardrum and other tiny parts of your ear start to move. This movement releases chemicals, and the chemicals give off electrical signals. These signals travel up the auditory nerves from your ears to your brain. When your brain receives the electrical signals, it translates them into information that you recognize as different sounds.

Sound energy is carried along in the form of waves. However, these waves don't go up and down like waves in the ocean. (A wave that bounces up and down is called a *transverse wave*.) Instead, sound waves make molecules bunch up and spread out in repeating patterns. When you make a sound, such as clapping your hands, the air molecules are pushed away from you in every direction. They form a bubble of high pressure around you, where the air molecules are squeezed together, and the bubble grows and grows as it moves. Behind it is a bubble of lower pressure that has fewer air molecules left in it. The waves continue to form as long as the sound continues. A wave formed by molecules that are squeezed together and pulled apart is called a *pressure wave* or a *longitudinal wave*.

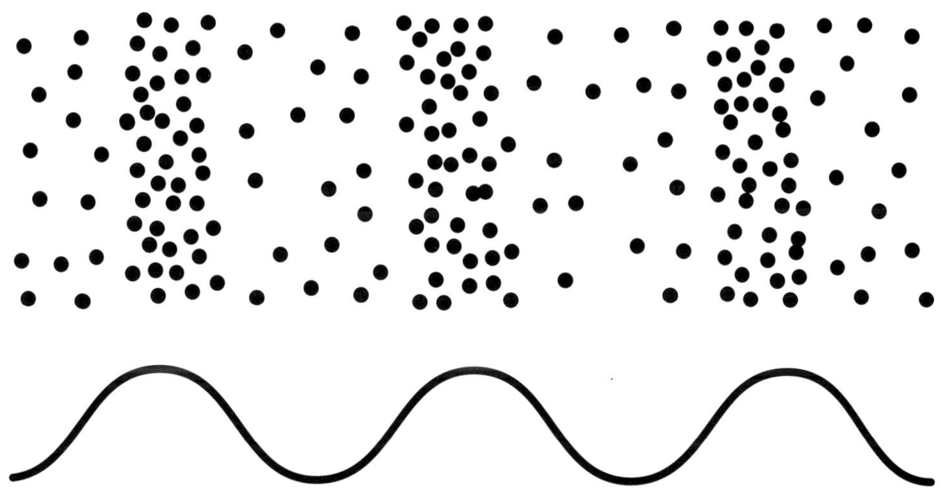

FIGURE I-2: **The dots at the top show how a longitudinal sound wave travels through the air. The dots represent molecules that are squeezed together and spread apart over and over. The line at the bottom is a waveform for the same sound wave. The "hills" represent the areas of high pressure, and the "valleys" represent the areas of low pressure.**

When studying sound waves, scientists use a drawing of a curvy line to help them understand what's going on. A diagram that shows areas of high pressure and low pressure as high and low points on a curvy line is called a *waveform*. You can use a waveform to help you measure different characteristics of a sound wave:

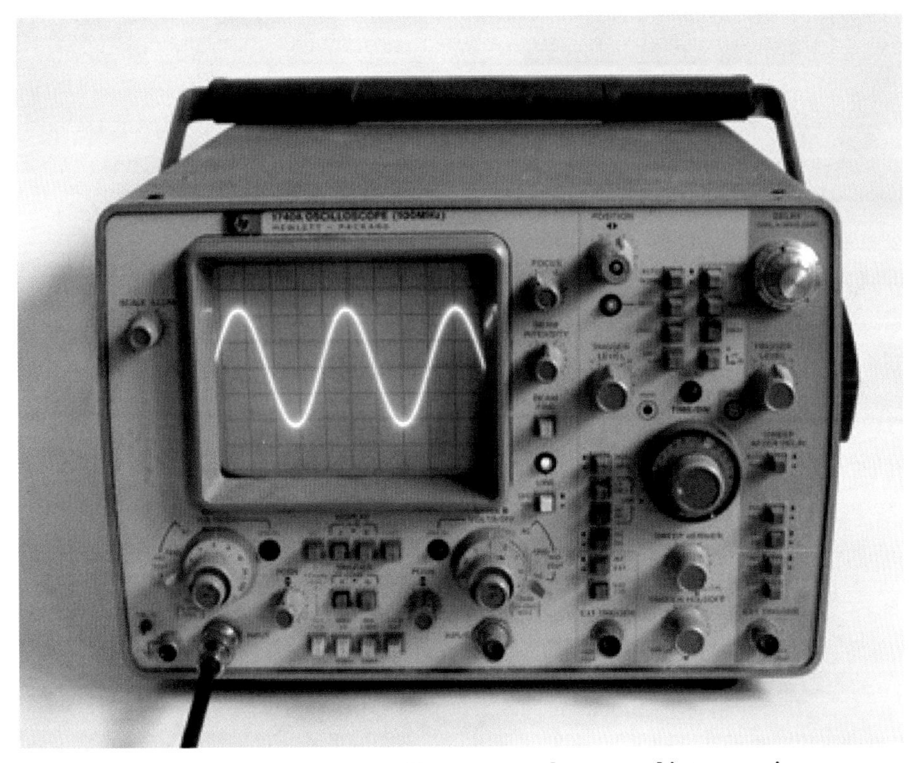

FIGURE I-3: A machine called an oscilloscope translates sound into a moving waveform image at the same time you hear it.

The *wavelength* is the distance from a particular point on one wave to the same point on the next wave.

Your measuring point can be the high point, low point, or midpoint of the wave, and it is measured in inches and feet or millimeters, centimeters, and meters. This distance makes up a complete wave, which is called a *cycle*.

The *frequency* of a wave is the number of complete waves that occur in a certain amount of time.

On a waveform drawing, you can tell the frequency by counting how many waves fit within a certain space. Frequency is measured in hertz. One hertz (written as 1 Hz) equals one cycle per second.

The *speed* of a wave is how fast it is traveling away from the point where it started.

The speed of a sound wave depends on the material it is moving through. In the air, the speed of sound is usually given as around 760 miles per hour (or 1,220 km per hour).

The *amplitude* of a waveform is the distance between its high point and its midpoint.

The bigger the amplitude, the more energy the wave contains. The amplitude of a sound wave is measured in decibels (dB).

THE CONNECTION BETWEEN WAVELENGTH AND FREQUENCY

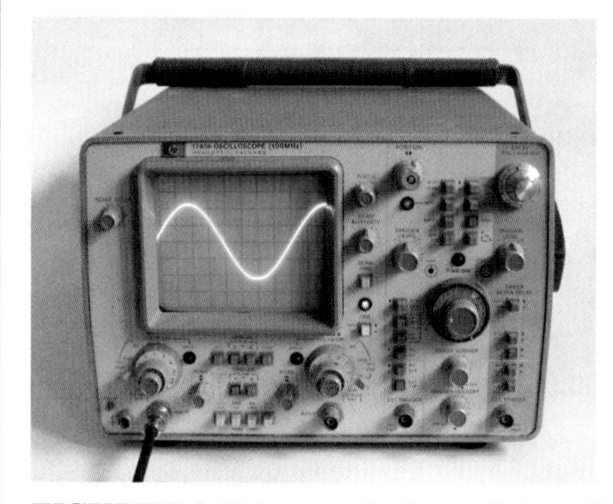

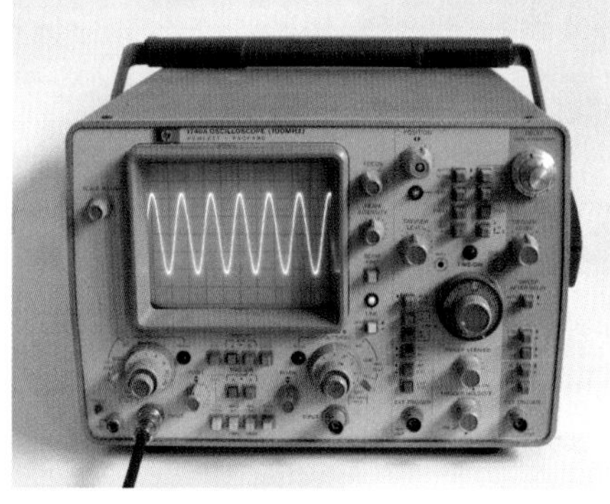

FIGURE I-4: **If the speed of sound stays the same, the number of complete waves that can fit within the period of one second (the frequency) depends upon how long the wave is (the wavelength). So longer wavelengths have lower frequencies (left), and shorter wavelengths have higher frequencies (right).**

Here's an important fact: wavelength and frequency are related! When one goes up, the other goes down. In other words, if you shorten the wavelength, you increase the frequency, because you can fit more cycles into the same amount of time.

In fact, if you know some math, you can convert one measure into the other. You'll also need to know the speed of the sound wave. To find the wavelength, you divide the speed by the frequency. To find the frequency, you divide the speed by the wavelength.

What Is Music?

All sound can be divided into two types: noise and music. But trying to define the difference between them can be tricky. Most definitions say that *music* is a series of sounds that have some kind of pattern, repetition, and meaning. *Noise* is made up of random sounds all mixed together.

Music doesn't have to sound pretty, but it must be created on purpose. That means a car horn honking, a hammer banging, or old-fashioned typewriter keys tapping can be considered music if they are part of a song—but not if you overhear them while going about your day. That can lead to interesting questions. Are the beautiful clear notes of tinkling wind chimes or splashing water drops creating music if they happen by accident, without any kind of pattern? What about a songbird that repeats the same string of notes over and over without "thinking" about it, or wolves that howl to communicate with each other? Animal sounds probably inspired the first humans to create music—can animals be musicians too?

In the end, the answer is up to you. When you create your own music, you can use any kind of sounds you like! But no matter what style of music you're playing, it helps to know a few basic terms:

A *note* is the smallest building block of music.

A note can be long or short, quick or slow, but as long as it lasts, the pitch (see the next definition) stays the same. Notes can be played one right after the other without a break, or they can be separated by long or short periods of silence, called *rests*. Notes are described by how high or low they are (see *pitch*), and how long they last (see the following *beat* definition).

***Pitch* is how high or low a note sounds.**

When someone asks what note you are playing, they really mean what pitch you are playing. In most cases, you can use either word to mean the same thing. The number of different pitches is limitless, but most music systems use between five and twelve notes.

A *beat* is a steady, repeating unit of time in a piece of music.

Like the ticking of a clock or the thumping of your heart, music stays at a certain pace (which can change as the song goes on). Written music is divided into sections called *measures* that contain a certain number of beats. The length of a note or a rest is also described in beats. Written music has a *time signature* that tells you two things: how many beats per measure (also known as the *meter*), and which note represents one beat. They are shown as two numbers, one above the other. In the music system used most often (sometimes called *Western music*), the time signature where a measure has four beats and a whole note lasts the whole measure is known as *common time*. In common time, a half note lasts half as long as a whole note, or two beats, and a quarter note lasts for one beat. So the time signature for common time is also called 4/4 time (which you say as "four four time").

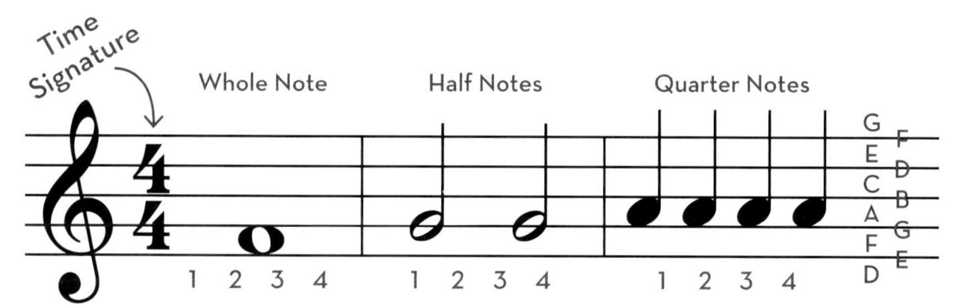

FIGURE 1-5: A music staff (set of lines) with a 4/4 time signature. It is divided into measures by vertical lines. Each measure contains four beats. You can tell the length of a note by whether it's filled in or open and whether it has a stem. You can tell the name of a note—its pitch—by which line or space it is placed on. The squiggly design on the left is called a G clef. It's a fancy letter G. The bottom part curls around the line for "G."

An *interval* in music is the distance between one pitch and another.

There is no limit to how close two notes can be, but different music systems pick different distances as their smallest interval. Intervals are measured in steps, and in Western music, the smallest interval between two notes is a half step. In other systems, intervals can be even smaller. The largest interval possible is the *octave*, which is the distance between the first and last note on an eight-note scale (see the following definition of *scale*). ("*Oct*" means "*eight*," like an eight-legged octopus.)

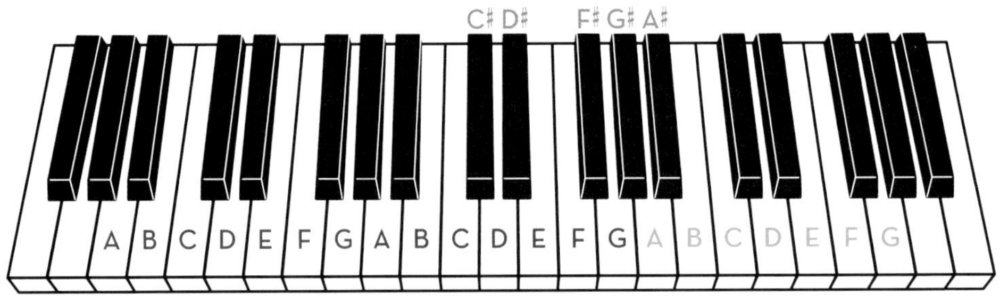

FIGURE I-6: This piano keyboard shows how the seven letter names given to the notes in an octave repeat over and over. The black keys on the keyboard are shown as sharp notes, but they could also be shown as flats. (For example, A# is the same note as B♭).

NAMING THE NOTES

How do musicians know what note to play or sing? They can tell by the name of the note and its position on the scale. A *scale* is series of notes following a specific pattern of intervals. There are a few different systems of names, and a few different types of scales, that are used in various parts of the world. Here are some important ones:

The Musical Alphabet

In Western music, there are seven notes with letter names (A, B, C, D, E, F, and G) that repeat as you go higher or lower. The interval between each two notes is either a half step or a whole step. You can see this repeating pattern on a piano keyboard. At some point in history, additional notes were added to the system for a total of 12 notes with equal intervals of one half-step between them. The extra notes are labeled sharp (♯) if they are higher than the letter and flat (♭) if they are lower. So F♯ is a half-step higher than a regular F (also known as F natural, written as F♮), and B♭ is a half-step lower than a B♮.

Chromatic Scale

Scales always begin and end on the same letter, with the first and last note one octave apart. (You learn why later in this introduction.) Most scales have eight notes—so a C scale goes C, D, E, F, G, A, B, C. When you include all the sharps and flats, you get a total of 12 different notes in an octave—so the chromatic scale is also known as the twelve-tone scale.

Pentatonic Scale

The *pentatonic* scale only uses five particular notes out of the possible eight notes in an octave. (*Pent* means *five*, like a five-sided pentagon.) The pentatonic scale is good to know, because any combination of notes always sounds "good," whether they are played one after another or at the same time. If you want to hear what it sounds like, try playing only the black keys on a piano!

Do, Re, Mi!

In some countries, the names Do, Re, Mi, Fa, Sol, La, and Ti are used instead of letters, with Do being the same note as C. In the United States, the names Do, Re, Mi, and so on are used to mean the notes in order, from low to high, no matter what letter note they represent.

The *timbre* of a note is the type of sound it makes.

Timbre is what gives different instruments their own character. Depending on its design, an instrument can produce a note that sounds crisp and clear, deep and rumbly, or scratchy and sharp. You can also get different kinds of timbre out of the same instrument by playing it differently—for instance, by plucking the strings of a violin instead of rubbing a bow across them.

Volume is how loud or soft a sound is.

Changing the volume up and down is a way to add meaning and interest to a piece of music. In written music, *dynamics* marks tell the player how loud or soft they should be playing in that part of the piece.

The Science of Sound and Music

If music is a kind of sound, and sound waves are a kind of energy you can measure and control, then changing the way an instrument produces sound waves should affect the kind of music it plays. Here's how you can use an understanding of the science of sound to make your music sound better:

The frequency of a sound wave is related to the pitch that you hear.

To change the pitch of a note produced by an instrument, you need to change the frequency of the part that is vibrating. And since frequency and wavelength are related, you can change the pitch of a note by changing the wavelength of the sound waves coming out of an instrument. In this book, you'll learn different ways to make notes sound higher or lower by fiddling around with the wavelength of the part of the instrument that vibrates.

Every instrument vibrates at its own special frequency or set of frequencies, depending on how it is made.

Whenever you give an object a push of some sort, it starts to vibrate. The rate at which an object starts to vibrate when it is disturbed is called its *natural frequency*—or for a musical instrument, its *harmonic*. The natural frequency of an object depends on things like its size and what it is made out of. The note that you hear when an instrument is vibrating at its lowest natural frequency is called the *fundamental tone*. When higher frequencies are also produced, these are called *overtones*. If the overtones are also harmonics (part of the set of natural frequencies), they can add richness and depth to the musical sound. If they are not harmonics, then the overtones can produce noise or buzzing on top of the musical tone that make the fundamental tone harder to hear.

Frequency and musical intervals are related.

When two notes have frequencies that are related mathematically, they sound better together than other combinations (at least to listeners of Western music). A *ratio* is used in math to show the size of one number compared to the other in the simplest form possible. For example, the frequency of middle C on a piano keyboard is roughly 260 Hz, and the

frequency of the C below it is about 130 Hz. Since 260 is twice as big as 130, we say the notes are in a ratio of 2:1 (read as "two to one"). Other intervals that sound good in Western music are the perfect fifth (with a ratio of 3:2), the perfect fourth (4:3), and the major third (5:4). When deciding what notes you want your instruments to play, keep these intervals in mind.

The timbre of an instrument depends on its natural frequency.

The way vibrations travel through an instrument produce its timbre. Simple sound waves create pure, clear notes. Instruments that produce combinations of sound waves with different frequencies have a richer, more interesting timbre. Again, different frequencies usually sound better together when they are related to each other by simple ratios. If the frequencies are unrelated, you end up with a jumble of sounds that register in your ears as noise rather than notes. As you work on the projects in this book, you can experiment with different designs and materials to see how they change the timbre of the instrument.

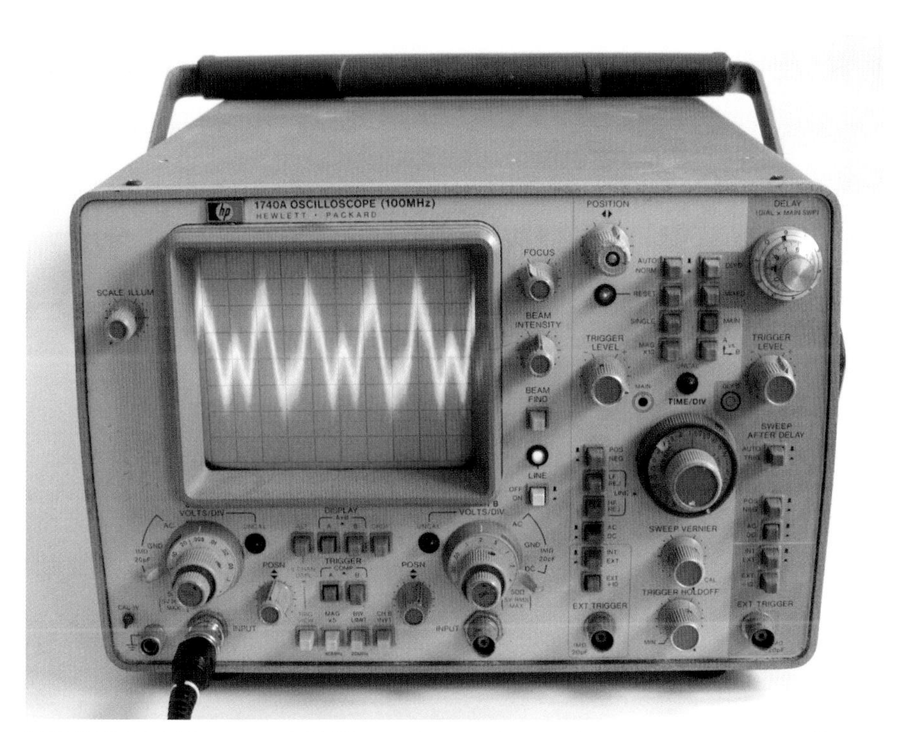

FIGURE I-7: When you add different frequencies together, you get more complicated waveforms.

The amplitude of a sound wave is related to the volume of the sound that you hear.

In other words, the higher the amplitude of a sound wave, the louder the volume. This makes sense, since a wave with a higher amplitude is carrying more energy. To make instruments sound louder, it helps to make other parts of the instrument all vibrate at the same frequency as the part creating the sound. When an object starts to vibrate because it is bombarded by vibrations, that is known as *resonance*. Instruments usually have resonators that make them sound louder.

Meet the Instrument Families

Musical instruments are usually divided according to the material they are made of and how they are played. In the standard orchestra, there are four families of instruments—strings (like violins and guitars), woodwinds (like clarinets and flutes), brass (like trumpets and trombones), and percussion (like drums and xylophones).

For designing instruments, however, it's more useful to describe instruments by the way they work. Every instrument (except electronic instruments, which we'll talk about in Chapter 4) has two important parts: the part that vibrates and creates the instrument's sound, and the resonator that helps make the sounds louder. The Sachs-Hornbostel system divides instruments into five families according to the way they vibrate and produce sound:

- **Chordophones**: Instruments with vibrating strings, including guitars and harps
- **Aerophones**: Instruments you play by making a tube of air vibrate, such as horns and flutes
- **Membranophones**: Instruments that have a thin skin that vibrates, like drums and kazoos
- **Idiophones**: Solid instruments where the entire body vibrates, such as bells and cymbals
- **Electrophones**: Synthesizers that use electronics and speakers to generate vibrations in the air

Now that you know a little more about the science and art of instrument design, it's time to make some yourself! In *Musical Inventions*, you'll see how instruments were made in the past and try out some new techniques using everyday materials. (Refer back to the "Musical Supply Closet" list on page xvi to find out what you need.) As you design and play your own instruments, you'll get an even better understanding of how sound waves and vibrations work—and what they have to do with good music.

Ready to create sounds no one has ever heard before? Let's get started!

1 Singing Strings and Warbling Winds

Put theory into action by building instruments that demonstrate how sound waves make music!

To really understand the connection between science and music, it helps to watch what happens when an instrument starts to shake, rattle, and roll. On some chordophones (string instruments), you can actually see the strings vibrate back and forth as they create sound waves, which is why it's so useful to use chordophones to study how music works.

Aerophones (wind instruments) are also good for demonstrating sound waves in action—even though you can't see air molecules vibrating. You can even change the length of some aerophones as you play, sliding between long and short sound waves and high and low notes.

In this chapter you make a variety of chordophones and aerophones and learn a little bit more about the science of sound waves. Try the projects and then use what you discover to invent some instruments of your own!

Playing Around with Notes

Instruments use different methods to change the pitch of the notes they play, and some use more than one method. To help you understand the science behind musical instrument design, here's a quick recap of what you learned in the Introduction.

- Every wave, including sound waves, can be described by three measurements:
  - Its wavelength (the distance from one point on the wave to the same point on the next wave)
  - Its frequency (how many complete waves pass by in one second)
  - Its speed (how fast the wave travels away from the point where the sound started)
- Frequency and wavelength are related—if the wavelength is shorter, more waves can pass by your ear in one second, so the frequency goes up.
- Frequency and speed are also related. The faster a wave is traveling, the more waves will pass by in one second, which also makes the frequency go up.

- The frequency of a sound wave is translated by your brain into a pitch (how high or low a note sounds). Every note has its own frequency, so changing the length or the speed of a sound wave (or both) will also change the pitch. On any instrument, the way to do this is by changing the part that's vibrating.
- Every instrument (or part of any instrument, like a guitar string) tends to vibrate at a particular frequency or frequencies, which are called its harmonics. The lowest note an instrument (or part of an instrument) can produce is its fundamental frequency, also called its first harmonic. Sometimes you will also hear higher frequencies, which are the instrument's overtones.

One interesting thing about instruments is that the sound you hear is actually made up of multiple sound waves blended together. Here's how it works: when a traveling wave hits a boundary between one kind of

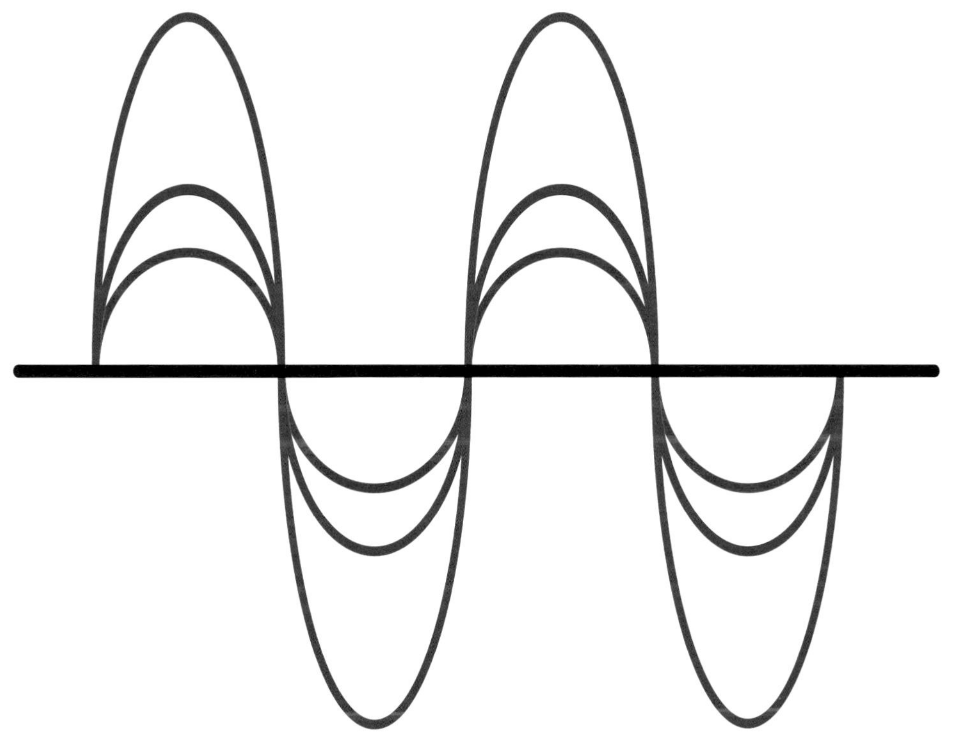

FIGURE 1-1: **When the blue wave and the red wave meet, the highest points and lowest points are added together. Together, they produce the purple wave.**

material and another, it bounces back. For example, when a sound wave moving down a vibrating string hits the end of the string is attached to, it turns around and heads back up the string. Soon you have sound waves with the same frequency traveling back and forth along the string in both directions at the same time. As the waves on the string overlap, they get added together. In scientific terms, when two or more waves cross over each other, that's called *interference*.

Where two high points or low points meet, you get points that move twice as high or low as the original waves.

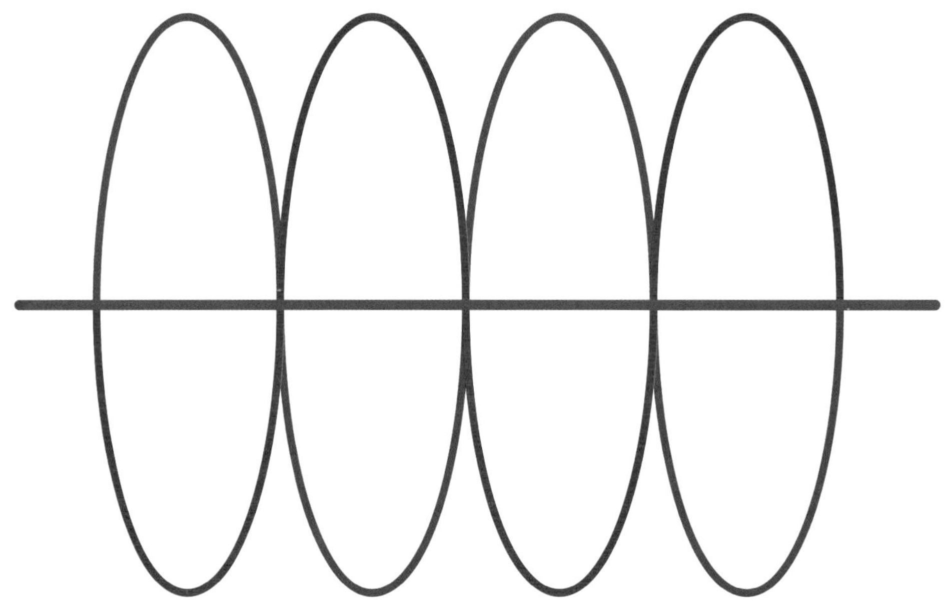

FIGURE 1-2: **When the blue wave and the red wave are equal, but the highest points and lowest points are opposite each other, the two waves cancel each other out. There is no movement up or down, just a straight purple line.**

Where a high point on one wave meets a low point on another wave, they cancel each other out, and you end up with a point that doesn't move at all. When the interference between two waves creates a wave that looks like it is no longer moving forward and back, you have a *standing wave pattern*.

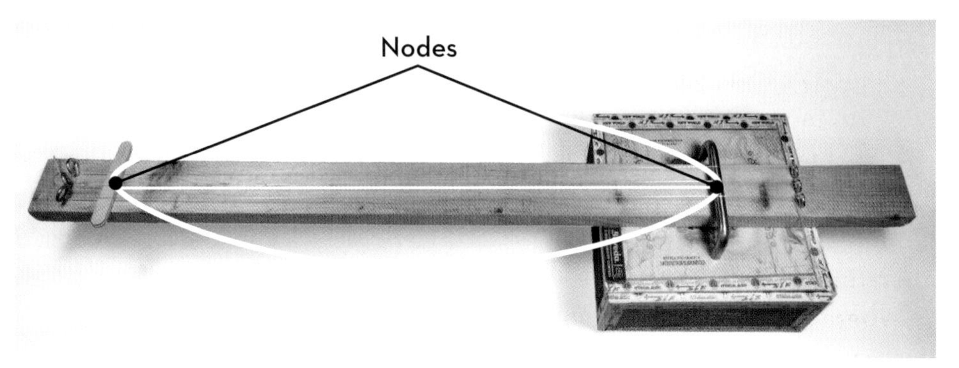

FIGURE 1-3: This standing wave pattern shows the first harmonic on the string of a cigar box guitar. There are two nodes that don't move, one at either end of the "loose" part of the string. The part of the string that swings back and forth like the curved white lines shown here is called the *antinode*. When you pluck the open string (without holding the string down with your finger), the note you hear is the fundamental frequency.

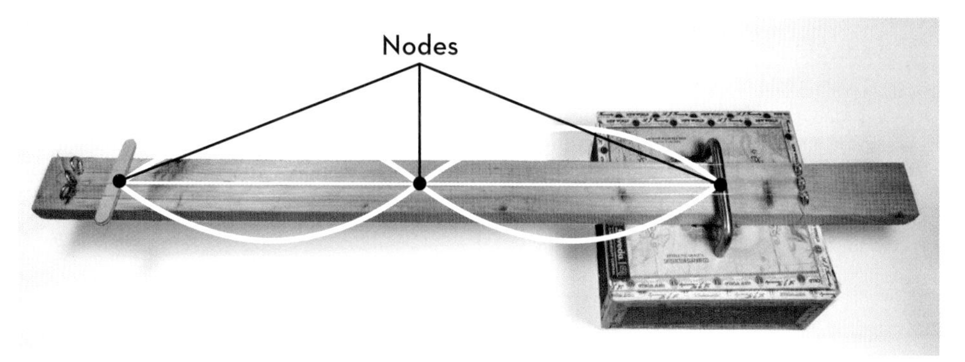

FIGURE 1-4: A standing wave pattern for the second harmonic. This has three nodes and two antinodes. If you pluck the string while touching it lightly at the node in the middle, you may hear the second harmonic. It is one octave higher than the fundamental frequency.

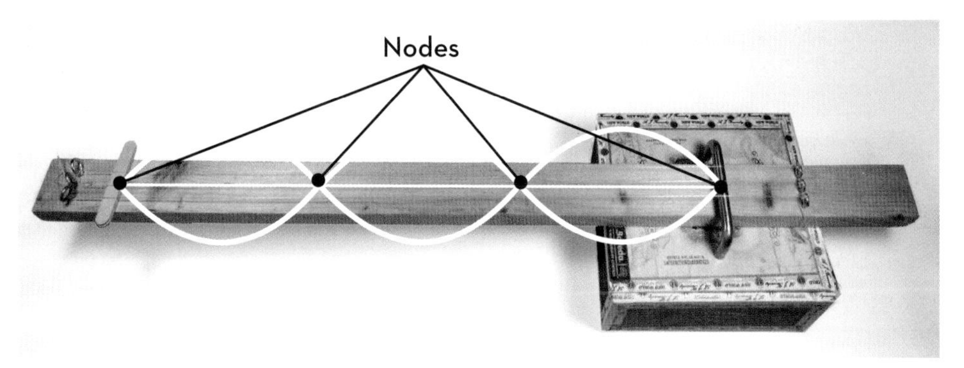

FIGURE 1-5: The third harmonic has four nodes and three antinodes. If you touch the string lightly at one of the nodes in the middle, you may hear the third harmonic. It is a fifth, or seven half steps, above the second harmonic.

The points that don't move on a standing wave pattern are called *nodes*. If you picture two people holding a jump rope, the jump rope represents a standing wave, and the two people holding the ends of the rope are the nodes. (The part that seems to go up and down on a standing wave pattern is called the *antinode*.) On any instrument, the first harmonic is the note produced by a standing wave pattern with two nodes.

The second harmonic is produced by a standing wave pattern with three nodes. In our jump rope model, it's as if you add a third person who holds the jump rope right in the middle, while the people at the end keep turning it. The two loose parts of the jump rope still go up and down, but each section is only half as long as before. For the third harmonic, you have four nodes that divide the vibrating part into three equal parts.

In Figures 1-3, 1-4, and 1-5, you may have noticed that the wavelength of a harmonic is related to the length of the string (that is, the part of the string that is free to vibrate). Figure 1-6 shows all three harmonics at the same time. The wavelength of the first harmonic is twice as long as the string. The wavelength of the second harmonic is the same length as the

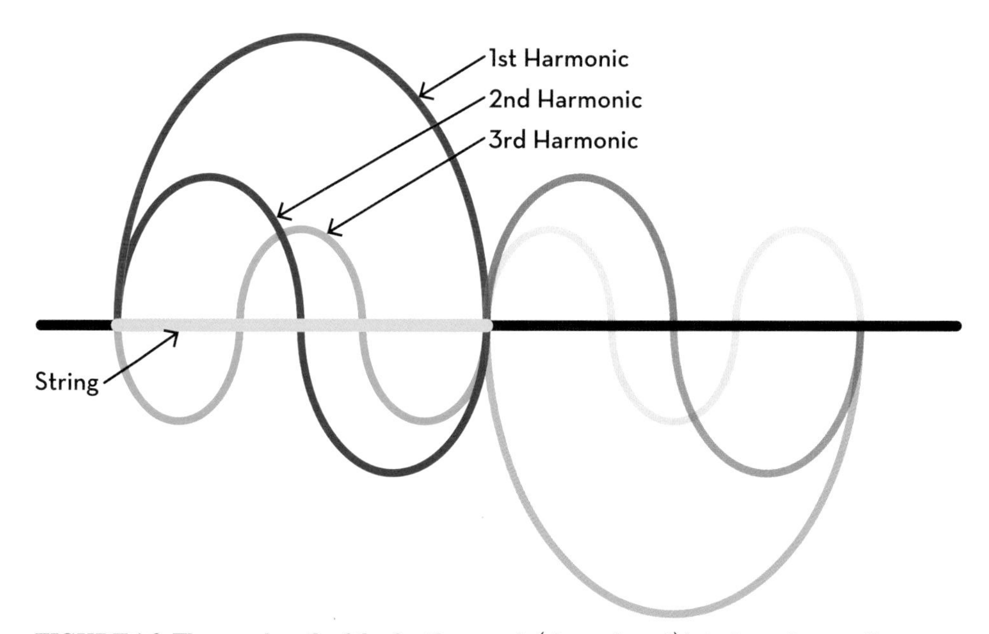

FIGURE 1-6: The wavelength of the first harmonic (shown in red) is twice as long as the string. The wavelength of the second harmonic (shown in blue) is the same length as the string. The wavelength of the third harmonic (shown in green) is $1\frac{1}{2}$ times as long as the string—if you doubled the length of the string, you could fit three wavelengths into it.

string. If you look closely, you can also see that the frequency of the second harmonic is twice the frequency of the first harmonic.

Remember the ratios you read about in the Introduction? This is where they come in. In math terms, we say the frequencies of the second harmonic and the first harmonic are in a ratio of 2:1. In musical terms, we say the second harmonic is an octave higher than the first harmonic. The ratio of the frequencies of the third harmonic compared to the second harmonic is 3:2. (In musical terms, the notes are a fifth apart.) So the reason why notes with frequencies in simple ratios like 2:1 and 3:2 sound good together is because they are created by the same vibration! (As you will see in Chapter 2, you can also have overtones that are nonharmonic.)

How to Change Notes on a Chordophone

Chordophones include all kinds of instruments with strings, from violins to guitars to ukuleles to harps. Even pianos and harpsichords (which have strings hidden inside them) are types of chordophones. On a chordophone, there are three main ways to control the frequency of the vibration produced by the strings, and they can be used separately or in combination:

Change the length of the string.

The longer the string, the lower the note it plays. Shorten the string and it will sound higher. You can change the length of the string while you're playing by pressing it down with your finger. The spot where your finger is pressing acts like the end of the string.

FIGURE 1-7: Inside a piano are strings. Each string is attached to a key on the keyboard. The strings are longer on the left side, where the keys play lower notes, and they get shorter and shorter as you move toward the right side of the piano, where the keys play higher notes.

Change how tight the string is.

When you pull a string tighter, it will play a higher note.

Loosen it and the note goes down. The measurement of how tight a string is stretched is its *tension*. Most chordophones let you tune the strings to the correct pitch with tuning pegs. When you wrap the end of the string around a tuning peg and turn it, you can increase or decrease the tension in the string.

FIGURE 1-8: The pegs on a violin let you tighten and loosen the strings to tune them.

Change how thick and heavy the string is.

If you have two strings that are the same length and tension but different thicknesses, the thicker string will play a lower note, and the thinner string will play a higher note. The thickness of a string on a chordophone is called its *gauge*. Most instruments have a standard length for the strings they use, so a string with a higher gauge will also have a higher mass. The *mass* is the amount of material in an object.

FIGURE 1-9: When you pluck a string on a bass violin, it vibrates at such a low frequency that you can see the blur of its movement.

How do tension and mass affect the pitch of a string? By changing the speed of the sound wave that is traveling along it. A sound wave will move faster along a string that's tighter or thinner than it will on a string that's looser or thicker. Since the speed is related to the frequency of the sound wave, that means that tighter or thinner strings have a higher fundamental frequency and play a higher note. The opposite is also true—that's why a string bass, which can play very deep low notes, uses strings that are thicker and looser than those on a violin or a guitar.

MUSICAL INVENTORS: PYTHAGORAS AND VINCENZO GALILEI

Some famous names in math and science invented rules to help us understand how string instruments work. The ancient Greek mathematician Pythagoras (who is better known for discovering the formula for finding the sides of a right-angle triangle) was the first person to describe the relationship between the length of a string and the note that it played in mathematical terms. And the effect of tension on the pitch of a string was discovered in Italy in the 1500s by Vincenzo Galilei, a musician who played the lute (an early type of guitar). Historians think he may have been helped by his son, Galileo Galilei—who launched a scientific revolution when he showed that Earth and the other planets all moved around the sun. Like his son, Vincenzo didn't just read about science and math. He did experiments to try to observe how things really worked. To figure out the relationship between tension and pitch, he hung different weights from strings to pull them tight by different amounts. He found that if two identical strings produced notes an octave apart, the higher string had four times the tension of the lower string.

Some chordophones only have one string, so it's easy to test the different ways of changing the pitch of a string instrument. Make this Compact Washtub Bass and try it out for yourself!

Project: Compact Washtub Bass

Materials

- Small metal pail (or large empty metal can, open at one end)
- Small eyebolt with two matching nuts
- Nail (about the same width as the eyebolt)
- Two large, flat washers with holes that fit the eyebolt
- Wooden screw-in extension handle for a paint roller (or a screw-in broom handle)
- Rubber cane tip
- Thin, soft nylon cord
- (Optional) large trapezoidal pencil eraser

Tools

- Hammer

FIGURE 1-10: This compact washtub bass is in the tradition of homemade jug band instruments.

A washtub bass is one of the homemade instruments that make up the jug band, a type of country music group popular in the early 1900s. It is usually made by connecting a piece of cord from the top of a wooden stick to the middle of a metal tub that has been turned upside down. The tub serves as a sound box, deepening and amplifying the sound. To play this instrument, you tilt the stick to make the string tighter or looser, creating higher and lower notes. Some players also use the stick as a fingerboard, pressing down on the string the same way you would with a guitar or violin. This

Compact Washtub Bass has a great sound and doesn't take up a lot of room. It's also just the right size for kids.

Safety Warning: Children should get an adult's help when they use hammers and nails. Also, be careful not to tilt the stick too far or the string may snap suddenly.

1. Turn the pail upside down, and use a hammer and nail to make a small starter hole in the center of the pail bottom.
2. Screw one of the nuts onto the eyebolt. Slide on one of the washers. Insert the eyebolt into the hole you made with your hammer and nail, loop side up. Inside the pail, slide the other washer onto the bolt. Fasten it on with the other nut.
3. Tie one end of the cord tightly to the loop in the eyebolt.

FIGURE 1-11: Make a starter hole for the eyebolt.

FIGURE 1-12: Insert the eyebolt into the hole in the pail, with the nut and washer attached

FIGURE 1-13: The inside of the pail, with the washer and nut holding the eyebolt in place

4. Put the rubber cane tip on the smooth end of the stick to keep it from slipping. Set the pail on the floor, and place the end of the stick with the rubber tip next to the eyebolt. Tie the other end of the string to the top of the stick, wrapping it around the threads so it doesn't slide down. The string should be tight.
5. Time to give the washtub bass a test! Put one foot on the edge of the pail to hold it in place. Move the bottom of the stick to the rim of the pail. The string should get a little tighter. Try plucking the string—do you hear a note? Now tilt the top of the stick away from the pail, keeping the bottom of the stick resting on the rim. The pitch of the string should get higher and higher.
6. To see how many notes you can play with your one-string bass, start with the stick as close to straight up and down as it can go and still play a loud, clear note. Tilt the stick a little bit to get the next note in the scale. (Think "do, re, mi" to get the notes of the scale.) Repeat this action to get the next few notes. At some point, you may have to

FIGURE 1-14: Tie the cord securely to the eyebolt.

FIGURE 1-15: A rubber cane tip fits right over the end of the stick.

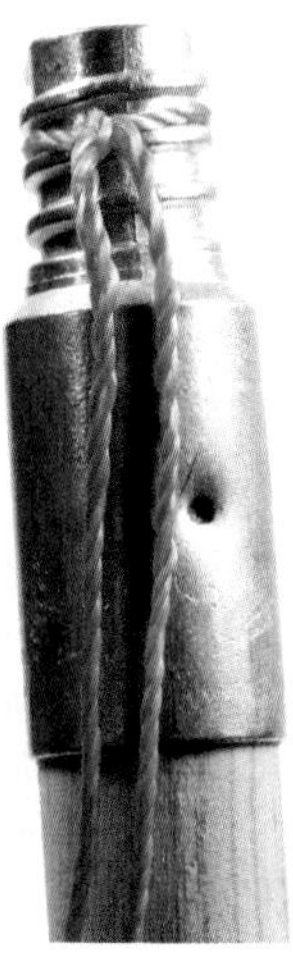

FIGURE 1-16: A paint roller stick has a threaded end that helps hold the string in place.

tilt the stick a lot to get the next few notes. But stop if the string starts to pull the bottom of the pail up—you don't want to break the string or rip the eyebolt out of the pail. If you don't quite get a full octave, try loosening the string a little bit and starting over.

7. To get more volume, try propping up one side of the pail with a flat pencil eraser. This lets the vibrating air out from under the pail and makes it easier for the sound waves to reach your ears.

FIGURE 1-18: Prop up the edge of the pail with a flat rubber eraser to let more of the sound out.

FIGURE 1-17: Just like on the bass violin, you can see the string on the washtub bass as it vibrates.

JUG BAND INSTRUMENTS

Create an entire jug band by gathering items from around the house or at the hardware store. Here's what you need:

Jug: The jug is the horn section of the band. If played right, it sounds like a buzzy trombone. Plastic maple syrup containers look the most like real ceramic jugs, but you can also try quart-sized milk or juice jugs. To play, hold the rim of the jug a little bit away from your mouth, purse your lips, and blow directly into the hole. Be prepared to make a rude noise, or even spit, to create the sound. Change notes by loosening or tightening your lips or by moving the jug closer or farther away.

Washboard: The washboard is a rhythm instrument. You play it by scraping something stiff against the ribs of the metal surface, such as a thimble or a whisk broom. Look for a replica washboard at old-timey souvenir shops, or try a ribbed paint roller tray or broiler pan.

Spoons: A pair of back-to-back spoons can also be used together as a rhythm instrument. Clack them together or against your knee. The trick is to grip the spoons firmly with the knuckle of your index finger between the handles, and your hand closed to make a fist. There should be a space between the bowls of the spoons. Stand with one foot up on a stool, and bang the spoons up and down between your thigh and the palm of your other hand.

Comb and Tissue Paper: This kazoo-like instrument creates a buzzing sound when you talk or sing. You need a comb with thin flexible teeth and a piece of tissue paper or wax paper the same width as the comb. Fold the paper in half and drape it over the comb. Let the paper hang loosely and put your mouth very close to it. Try saying a sound like "do do do" until you feel the paper vibrate. Once you've got the hang of it, you can sing notes and use different syllables to change the sound.

Project: Stick-on-Top 3-String Strummer

Materials

Cigar box, wooden clementine box, cookie tin, sturdy cardboard box, plastic orange juice jug, or other container for the body

1×2 strip of wood, about 3 feet (1 m) long

2–4 round-head wood screws, ¾ inch (1 cm) long, with washers to fit (A good size is #8-32.)

6 small eye screws for tuners (¼ or ⅜ inch [roughly 6 or 9 mm] are good.)

2 mini wooden craft sticks or a small wooden dowel

White multipurpose glue (or wood glue)

1 metal drawer pull with screws (the kind shaped like a squared-off "U"), wide enough to fit over the neck and smooth on top (You may need to replace the screws that come with the drawer pull with shorter screws; they should be about ½ to ¾ inch [roughly 1 cm] long.)

Scrap lumber (for placing under the cigar box while you are banging in nails)

Nylon fishing line—about 3-foot (1-m) lengths of three different weights, such as 20, 30, and 50 pound (9.1, 13.6, and 22.7 kg)

Tools

(Optional) sandpaper

Pencil

Masking tape

Ruler

Screwdriver

Drill or hammer and nail (The drill bit or nail should be thinner than the wood screws.)

There aren't a lot of parts to this primitive guitar-like instrument: a neck to hold the strings, a body to act as a sound box to amplify the vibrations, screws to let you tighten the strings to the notes you want, and a nut and a bridge to hold the strings a little away from the neck at either end. Depending on how many strings you add and how you tune them, your strummer can be used like a traditional cigar box guitar (or CBG for short), a slide guitar, a ukulele, a dulcimer, or another type of chordophone. Or just invent your own kind of stringed instrument!

FIGURE 1-19: A cigar box guitar is a classic DIY instrument.

1. If the wooden strip you plan to use for the neck feels rough, use sandpaper to smooth it on all surfaces. If you want to finish it, you can paint the wood or seal it with linseed oil or a stain.
2. Next, attach the neck to the body of the instrument. If you are using a cigar box, open the lid and hold it straight up. On the inside of the lid, measure and mark the center along the top edge. Stand the neck up against the outside the lid. Trace a line on the neck and measure and mark the center. Make sure the two center marks are lined up. Use

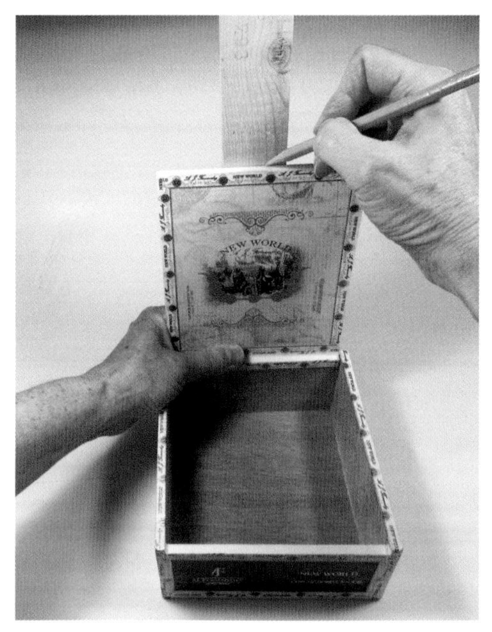

FIGURE 1-20: Draw a line to show where the neck will attach to the lid of the cigar box.

masking tape to attach the neck temporarily.

3. Next, lay the neck down on a sturdy work surface with the cigar box lid on top of it. Put pencil marks where you want the screws to go. It's best to zigzag them rather than line them up one above the other. Use a drill (or a hammer and nail) to make starter holes for the screws. Then screw in the screws through the box lid and into the stick.

4. Close the box so the neck is now on top of the lid. Near the end of the stick closest to the box, draw a line straight across. This is where the bottom screws will be attached. Mark one spot for each eye screw—one for each string. They should be evenly spaced and far enough apart that you can twist the loops without them bumping into each other. Do the same thing near the other end of the neck, at the top of the fingerboard, furthest from the box. Screw in the eye screws. If you have trouble turning the screws, you can insert a thin screwdriver into the eyehole as a handle.

FIGURE 1-21: **Screw the lid to the neck.**

FIGURE 1-22: **Avoid putting the screws in a straight line.**

FIGURE 1-23: **Leave space between the eye screws so they can turn without bumping into each other.**

5. Take the drawer pull and place it over the stick. This will be the bridge that holds the strings above the body. With the pencil, mark where the ends attach to the box. Measure the distance between them to be sure it is accurate. Also measure the distance from each mark to the bottom of the box to make sure they are even. Then place the lid over a piece of scrap wood and carefully make a starter hole straight through the lid at each mark. (The scrap wood supports the thin lid and protects the surface below it as you drive in a nail or a drill bit.) Then from the inside of the lid, screw in the screws to attach the ends of the drawer pull securely.

6. Now attach the strings one at a time. The lowest note usually goes on the left, so start on the left with the thickest string. (If you are left-handed, you can string it the opposite way.) Tie one end to the bottom eye screw furthest left with a double knot. Tie the other end to the top. Twist the screw at the top until you get a good sound. Repeat with the next thickest string. Tune it to an interval that sounds good

FIGURE 1-24: Use a screwdriver as a handle to help you turn the eye screws.

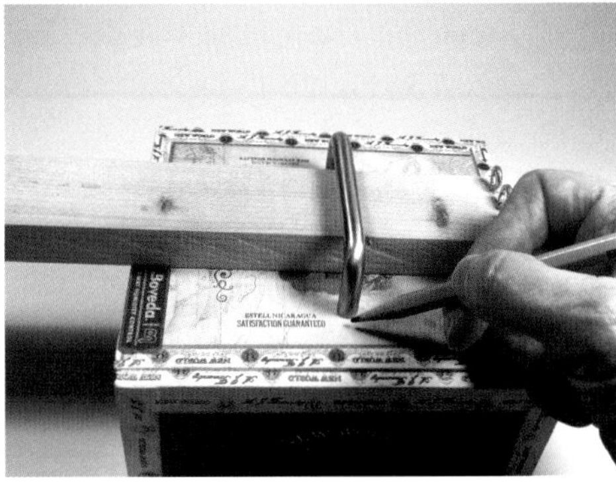

FIGURE 1-25: Mark the lid to show where the drawer pull will be attached.

FIGURE 1-26: Make starter holes for the drawer pull screws.

when you play the two open strings together. An *open string* is one that you play without pressing it down to shorten it. Continue with the remaining strings until you have them all attached. (See the sidebar titled "Tuning Your Strummer" later in this chapter for some suggestions on tuning your instrument.)

7. To reduce buzzing when you play the strings, add a smaller version of a bridge at the top of the neck to hold the strings a little above the fingerboard. This piece is known as a nut. You can make a nut using a small wooden dowel or two mini craft sticks glued together. Slide the nut under the strings across the neck, near the eye screw tuners. Adjust it until it gives you the best possible sound, and mark the spot. Be careful to place the nut close to the screws, to prevent it from getting in the way of your fingering when you play. To glue the nut on without removing the strings, loosen the strings slightly and slide a pencil under the strings to hold them away from the neck. Squeeze a very thin line of glue onto the nut, and then

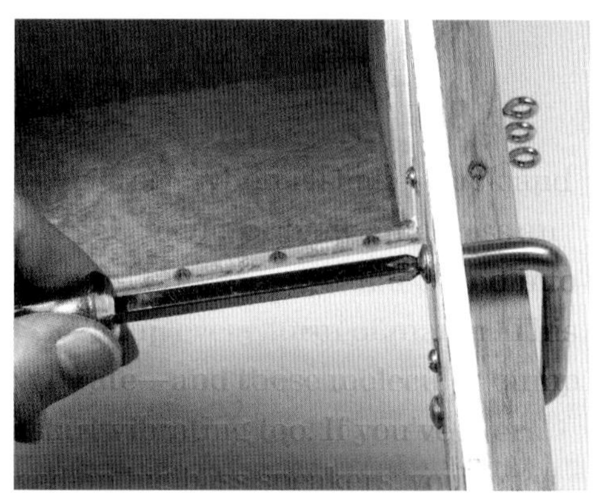

FIGURE 1-27: Attach the drawer pull to make the bridge.

FIGURE 1-28: Tie the strings tightly.

FIGURE 1-29: Make sure to double knot your strings.

carefully slide it into place. Wipe up any excess glue. Remove the pencil. Use clamps, large binder clips, or crisscrossed rubber bands to hold the nut tight against the neck until the glue is dry.

Adaptations, variations, and extensions:

- For a narrower neck (good for smaller hands), use only two strings instead of three.
- If your instrument's neck is wide enough, use larger eye screws to make them easier to turn and to tune. You can also zigzag them rather than place them in a straight line.
- Use four strings instead of three to turn your instrument into a ukulele.
- *Frets* are small ridges that make it easier to press down the strings to get different notes. Some CBG makers simply draw on frets with permanent markers to show where your fingers should go. You can also make frets by gluing wooden toothpicks or pieces of bamboo skewers at the fret marks. (Doing this may be easier if you temporarily remove the strings.) You can make a sound hole (about the size of a quarter) in the lid next

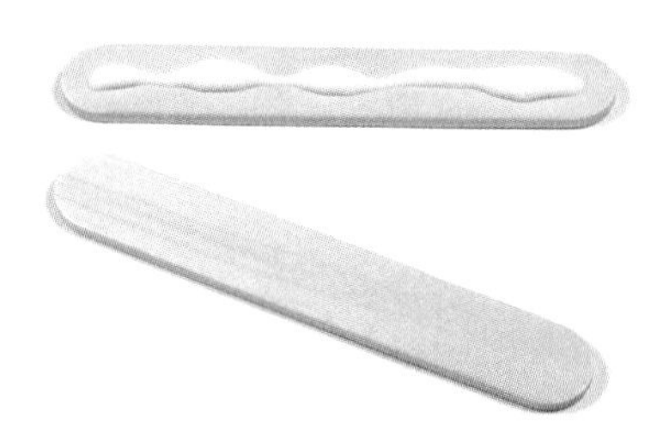

FIGURE 1-30: **Glue two mini craft sticks together to make a nut.**

FIGURE 1-31: **Rubber bands hold the nut in place on the neck while the glue dries.**

FIGURE 1-32: **The neck with the nut attached**

to the string to release more of the sound from inside.

- Make a slide that lets you press down all the strings at once. A traditional slide is made from the neck of a glass bottle, cut off and smoothed. You put your finger inside the slide and rub it up and down the strings to create different types of sound. You can make your own slide from an old glue stick cap or other small sturdy plastic tube. If it's too small to slip over your finger easily, just hold it against the strings.

FIGURE 1-33: **Cigar box guitars are a good size for young players.**

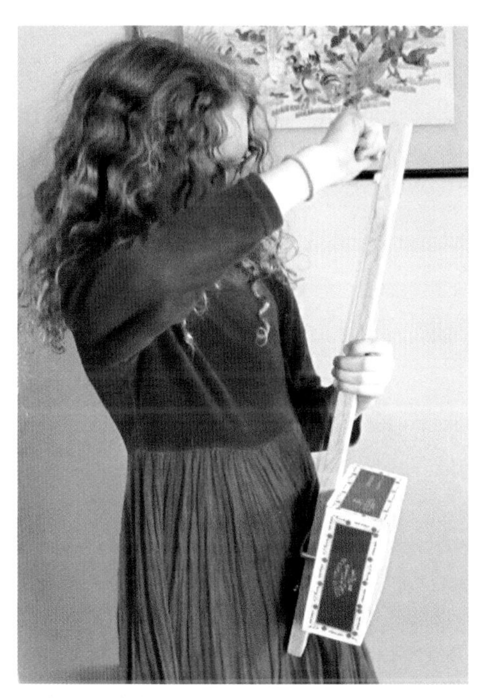

FIGURE 1-35: **Tune your guitar by turning the eyehole screws at the top of the neck.**

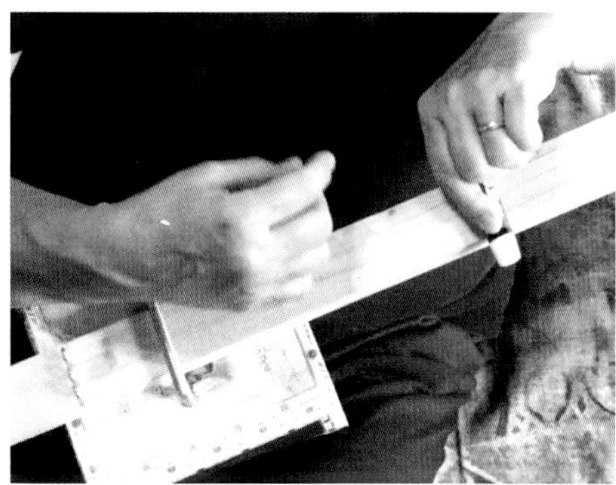

FIGURE 1-34: **Press a smooth, tube-shaped object, like an empty glue stick tube, onto the strings as a slide.**

TUNING YOUR STRUMMER

One popular tuning for three-stringed cigar box guitars is C-G-C, or other notes where the outside strings are an octave apart and the middle string is five notes above the lowest note. You don't have to worry about tuning to actual notes unless you want to play along with other instruments or recordings.

Another option is to tune the strings so they are each a third apart. (In Do-Re-Mi terms, if the first string is Do, the second string would be Mi, and the third string would be Sol.) Either way, you will be able to strum an open chord across all the strings.

How to Change Notes on an Aerophone

Aerophones include all kinds of instruments where the part that vibrates is the air in or around them. They include wind instruments that you blow into, like recorders. On some, like flutes and panpipes, you start the air vibrating by

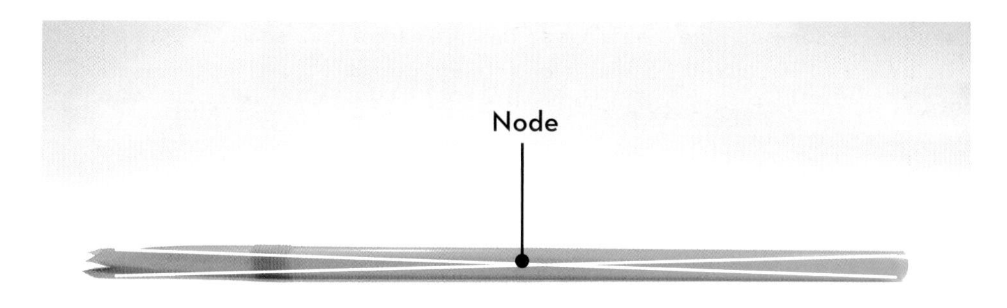

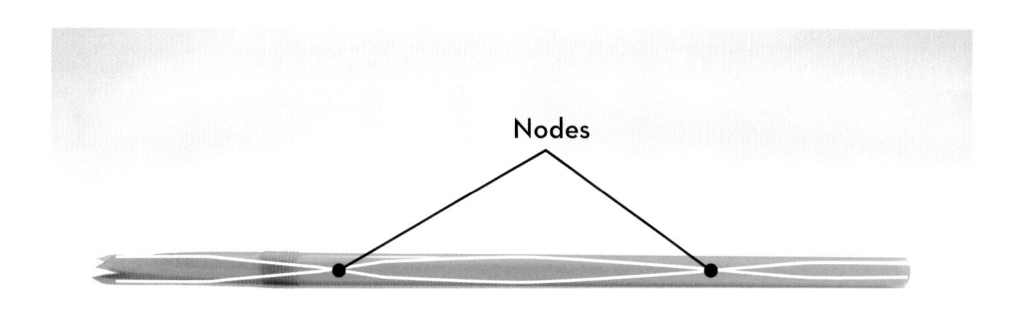

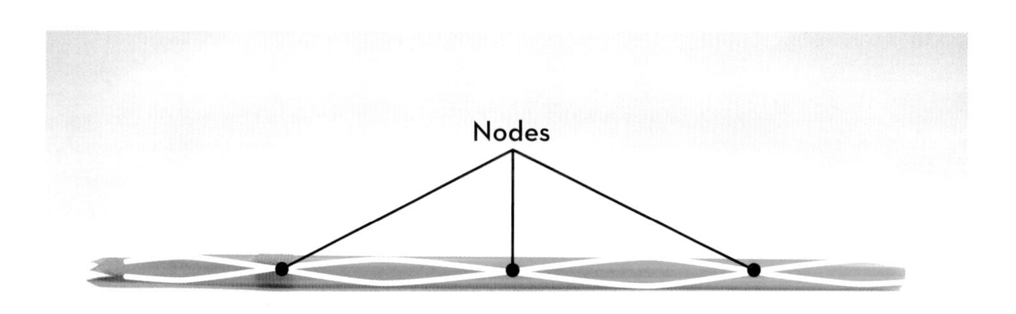

FIGURE 1-36: The first, second, and third harmonics in an aerophone with both ends open

blowing across an opening. Other aerophones, such as the trumpet, tuba, and didgeridoo (a very long pipe from Australia, usually made from a hollow eucalyptus tree) require you to buzz your lips while you blow. Many wind instruments use a *reed*, a thin piece of material you hold in your mouth to create the vibration. These include the saxophone and clarinet, which use a single reed, and the oboe and the bassoon, which use a double reed. But some aerophones don't need you to blow into them at all. Accordions and pipe organs use a bellows that is squeezed or pumped up and down, and calliopes use steam!

On a chordophone, it's easy to see the part that's vibrating—the string. With an aerophone, the part that's vibrating, the air, is invisible. But just like with string instruments, the length of a wind instrument tells you what the wavelength of its sound waves will be. For tube-shaped instruments, such as the flute, the ends of the tube act like the ends of the string. Even if the tube is open at one or both ends, the change in air pressure between the inside and the outside of the tube is enough of a barrier to cause the sound wave to reflect back and create a standing wave pattern.

One way to play different notes on a tube-like aerophone is to add holes. When the holes are open, they act like the end of the tube to make the sound wave reflect back. When the holes are closed, the sound wave keeps going as if the hole isn't there. Other aerophones, such as trombones, use extra pieces of tube that can slide in and out to make the tube longer or shorter. And changing the shape or position of your mouth can also change the sound wave the aerophone produces.

As you try these aerophone projects, see if you can find ways to change the notes they play. Some of the projects also let you hear the different harmonics an instrument can make. You may hear them created one at a time, or all together, as very faint overtones.

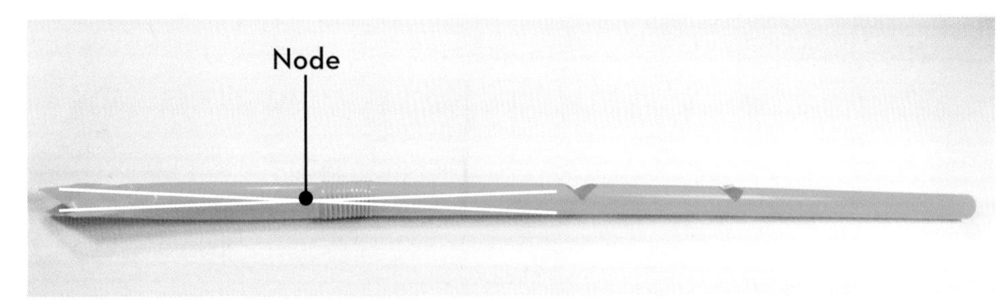

FIGURE 1-37: When the finger holes on a tube-shaped aerophone are open, the standing wave ends at the open hole. What do you think happens if there is more than one hole open?

Project: Drinking Straw Aerophones

Materials

- Plastic drinking straws, preferably bendy, plus extra straws
- Scissors
- Tape (the wider the better; masking tape is a good choice)
- (Optional) permanent marker
- (Optional) hole puncher

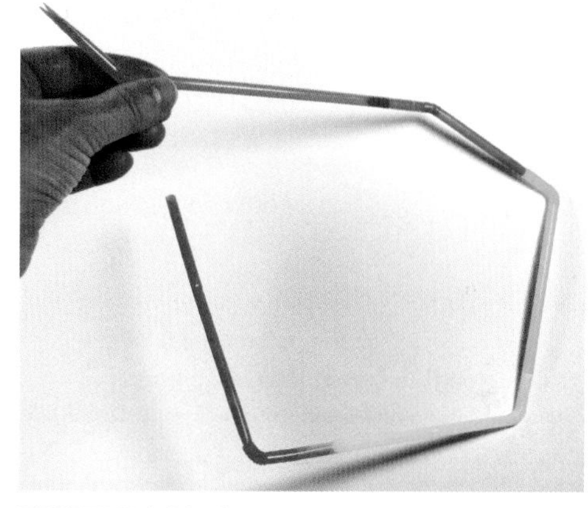

FIGURE 1-38: An aerophone can be as simple as a drinking straw.

Turn a drinking straw into an instrument by giving it a double reed like an oboe. Then use the science of standing waves to create drinking straw aerophones that can play a range of notes.

1. Bite down on one end of the straw to make it flat. Cut the flattened end to make a V shape. Open the flattened part of the straw a little.

FIGURE 1-39: Bite on the end of the straw to flatten it.

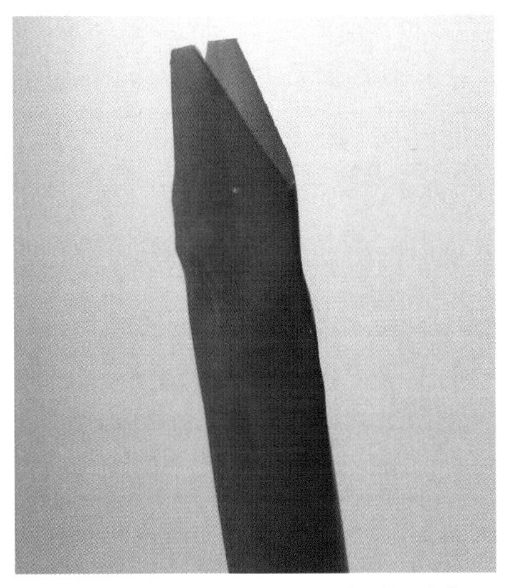

FIGURE 1-40: **You can trim the tip of the straw if you don't want it to be so pointy.**

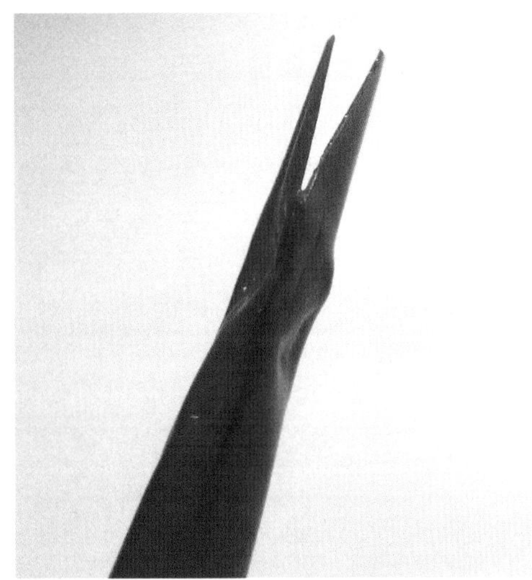

FIGURE 1-41: **After you cut the end into a V shape, open the flaps a little to create your double reed.**

2. To play, put the V in your mouth and blow.
3. Make another straw and blow on it the same way, but this time, snip off a piece while you're blowing. You'll hear the pitch get higher as the column of air inside the straw gets shorter. Keep cutting and you'll hear the note go up and up. How short can your straw get and still produce a sound?

FIGURE 1-42: **You may have to blow hard to get a sound.**

Variations:

♫ Add finger holes to create different notes without changing the length. Measure the length of the straw, and then use a permanent marker to put dots halfway from the end, one third of the way from the end, and so on. Pinch the straw and cut a small V-shaped opening at each mark. (You can also use a hole puncher to make the holes perfectly round. Now cover the holes with your fingers as you blow. As you uncover each one, the pitch should change to notes related to the pitch of the whole straw.

FIGURE 1-43: **Every time you cut the straw the notes gets higher!**

- Try making an extra-long straw oboe by connecting additional straws. Take a new straw or piece of straw and squish it enough to slide it inside the end of the oboe. Keep adding more pieces to see what notes you can produce. If you have bendy straws, you can bend the extra pieces into interesting shapes. Does bending the straw affect the sound?
- If you have two straws that are slightly different widths, try making them into a trombone. Slide the fatter straw over the thinner straw. Make a reed on the thinner straw, then slide the wider straw back and forth to change the pitch by making the instrument longer and shorter.

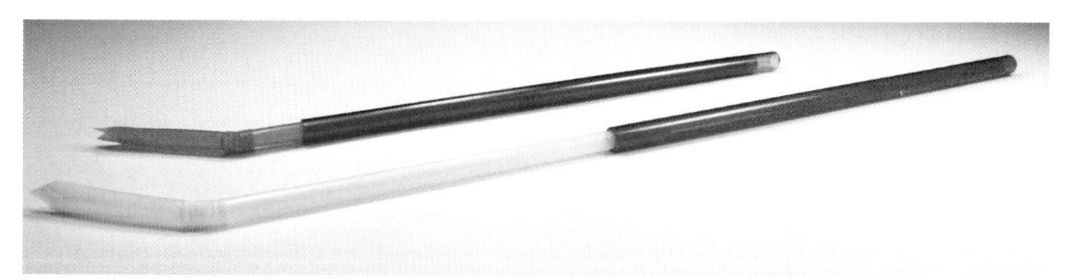

FIGURE 1-44: **If you have a straw a little wider than your oboe, you can turn it into a double-reed trombone! Slide the outer straw back and forth to change the pitch.**

Project: Giant Corrugated Singing Tube

Materials

Lightweight plastic corrugated hose (such as drainage hose from the plumbing aisle), about 4–6 feet (1.5–2 m) long

Tools

Heavy-duty scissors or utility knife

FIGURE 1-45: **Find a place to play the corrugated tube where you won't hit anything or anyone.**

You may have seen singing tubes in the toy store. When you swing them around, you force air to rush through them. The bumpy insides churn up the air and create sound waves. The faster you swing the tube around, the faster the speed of the sound waves and the higher the pitch.

One unique feature of the singing tube is that it can only play certain notes—those related to the fundamental frequency. You'll never get a note that's out of tune, because the note jumps directly from one to the next. If you want to play notes other than those that are harmonics of the fundamental frequency, you'll have to get a tube with a different length or width.

Safety Warning: Only play the giant corrugated singing tube in an open space—preferably outdoors—where it can't hit furniture, walls, or other people. Young musicians may need adult help to cut the tube.

1. If you need to cut the hose to the proper length, use heavy-duty scissors or a utility knife to make a neat cut. Use the bumps as a guide.
2. To play the tube, find an open space with plenty of room around you. If it's long enough, hold one end of the tube pointing toward your ear. This is the end the sound comes out of. With your other hand, grab the middle of the tube. The loose end is what you will swing, so make it short enough to be comfortable.
3. Begin to swing the tube around. As you twirl it faster and slower, listen closely to find the speeds that produce the clearest, loudest sounds. You may also be able to hear multiple overtones playing at the same time. See how many different notes you can produce with one tube!

Variations:

- If you have extra tubing, try cutting different lengths to hear the different notes they play. Try lengths that are twice as long, or one-third as long, as your first tube. Count the corrugations (the bumps) as a guide.

FIGURE 1-46: A wider hose produces lower notes.

FIGURE 1-47: The hose is easy to cut with scissors or a utility knife.

FIGURE 1-48: The hose should be long enough to swing around and force air through it.

- Keep an eye out for other widths of tubing at hardware stores and elsewhere and create a corrugated tube orchestra!

FIGURE 1-49: **If it's long enough, point one end of the hose toward your ear (or your audience) and swing it from the middle.**

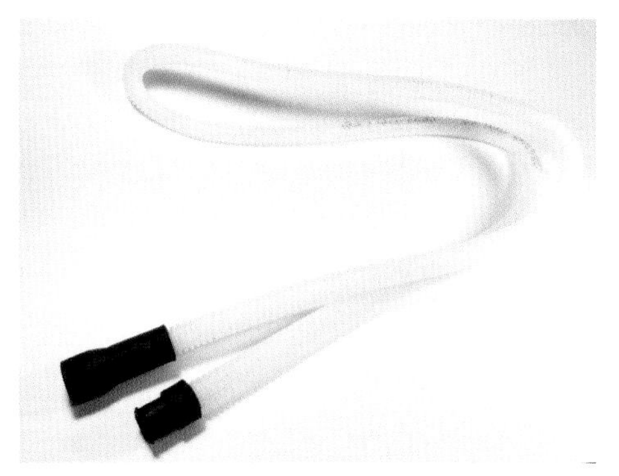

FIGURE 1-50: **Different widths produce different notes. This long hose produces several tones.**

MUSICAL INVENTORS: P.D.Q. BACH

Although the singing tube is a toy, corrugated tubes have been used in classical music. The composer Peter Schickele, who also goes by the name P.D.Q. Bach, wrote several fun pieces that included a vacuum cleaner hose, which he called a *lasso d'amore*. (A *lasso* is a rope with a circle tied in the end, twirled by cowboys as entertainment and to capture cattle. *D'amore* is Italian for "love.") Another P.D.Q. Bach piece, *Eine Kliene Kiddiemusik,* has parts for three musicians who must hit themselves on the head with plastic tubes.

Project: Bullroarer

Materials

- Wooden paint stirrer (sometimes free in paint or hardware stores)
- Sandpaper, coarse and medium grit
- Thin nylon string, about 3 feet (1 m) long
- (Optional) markers, paint, or wood stain
- (Optional) work gloves, safety glasses

The bullroarer is an ancient kind of instrument that you swing around on a string. As it twirls, it produces a low humming sound that can be kind of eerie. The bullroarer has been used in religious ceremonies in different parts of the world, including Europe, North America, and Australia. It is usually made of wood, clay, or bone.

You may think that having a string means a bullroarer is a type of chordophone. In fact, it is a *free aerophone*—an instrument that creates sound waves by cutting through the air around it. As you swing the bullroarer around, it twists around on its string. This twisting stirs up the air and creates the sound waves.

You can change the sound of a bullroarer by spinning it faster or slower. In fact, you may hear the note it plays go up and down with each swing as your arm speeds up and slows down. Different sized bullroarers will also create different sounds. Try making one yourself and then see what you can do to vary the notes it plays!

Safety Warning: Only play with the bullroarer in an open space—preferably outdoors—where it can't hit furniture, walls, or other people. You may want to wear safety glasses when you play it. Work gloves will protect your fingers from getting blisters as you twirl the string.

1. Most bullroarers are shaped like airplane wings, with rounded edges that are thinner than the center of the weight. Take the coarse sandpaper and sand down the edges of the paint stirrer to make them thinner and rounder. Use the medium sandpaper to make the surface smoother. Clean off any sawdust. If you want, you can decorate your bullroarer with markers, paint, or wood stain.
2. Attach the string by tying it tightly around the indent in the paint stirrer. Keep tying knots to make a little "neck" that will let the stick twist as you twirl it around on the string.
3. Take your bullroarer to a wide open spot to test it out. Make sure that no one is close enough to be hit when you start spinning it around (or if it goes flying off accidentally). Wrap the string around your hand until the string is a comfortable length. Let the bullroarer hang down. Then twist the bullroarer at the end of the string as many times as possible. The string should coil up slightly. Start to swing it around over your head, or next

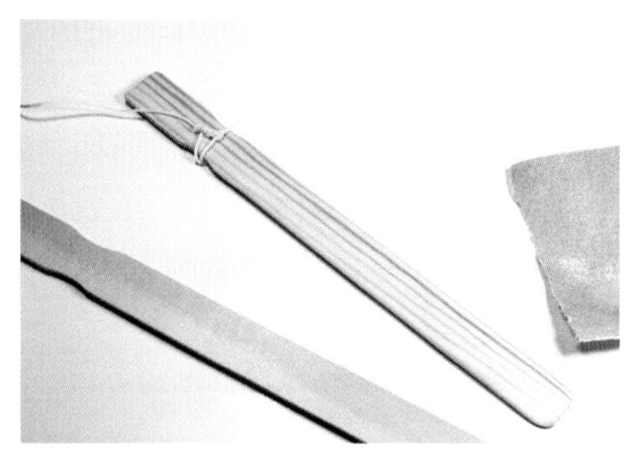

FIGURE 1-51: **Wooden paint stirrers are easy to sand to an aerodynamic shape.**

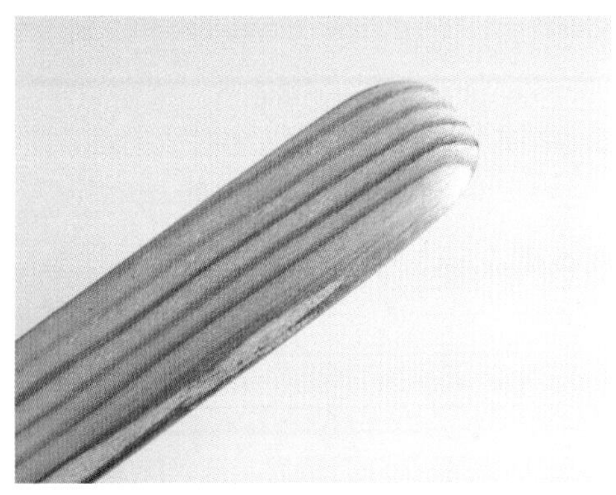

FIGURE 1-52: **Sand down the end of the paint stirrer until it is tapered to a thin rounded edge, like an airplane wing.**

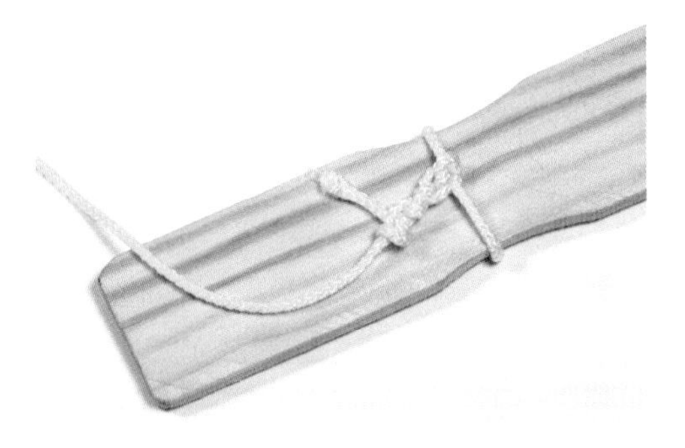

FIGURE 1-53: **Tie a row of knots to make it easier for the stick to swivel around.**

to you, like a cowboy lasso. You should hear a low humming or buzzing sound. If the bullroarer doesn't twist well and doesn't create a sound, try a thinner string, such as kite string.

Extensions:

- Try making bullroarers that are heavier or lighter to see if the different masses play different notes. For instance, you could glue two paint stirrers together, or tie wooden beads onto the string.
- What other ways can you change the weight of the bullroarer?

FIGURE 1-54: This photo shows how the bullroarer unwinds after you twist up the string. As you swing the bullroarer around in a big circle, it should also be twisting and untwisting to create sound waves in the air.

FIGURE 1-55: Find an open area to test your bullroarer.

Project: The Pickle-O, a Vegetable Ocarina

Materials

Small cucumber, zucchini, sweet potato, regular potato, or large carrot (about 5 inches [12 cm] long is good)

Apple corer, or a vegetable peeler with a corer top

Plastic or rounded dinner knife

Round toothpicks

An *ocarina* is also called a *vessel flute*. Instead of an open-ended tube, it is more like a miniature clay jug with holes. Ancient versions were often made in the shape of animals or people.

The ocarina works like a whistle, but it has finger holes that let you play different notes. The modern ocarina was invented in the late 1800s by Giuseppe Donati. Legend has it that Donati was a brickmaker who saw an ocarina from South America in a concert and made his own by firing clay models in his oven. He standardized the holes so that the ocarina could play along with other modern instruments.

The most common type of ocarina is sometimes called a sweet potato ocarina because of its shape—but you can make actual ocarinas from many kinds of vegetables! Cucumbers are a good size and soft enough to carve easily. They're juicier than other vegetables, though, so you'll need to dry them off as you work.

Your vegetable ocarinas may not last long, but they are fun to make and to play. This project was inspired by YouTube videos by the Vegetable Orchestra of Vienna, musician Junji Koyama, and a video tutorial by Dave Hax.

FIGURE 1-56: A cucumber makes a tasty wind instrument.

Safety Warning: Watch out for slips with knives and other kitchen tools when cutting or peeling vegetables. Kids should get adult help.

Note: A power drill with large and small drill bits can be used to hollow out the vegetable and to make finger holes. (Kids should get adult help to use power tools.) Drill bits (or screwdrivers) can also be turned by hand to carve out vegetables.

1. Thoroughly wash the skin of your vegetable.
2. To make your vegetable into an instrument, you need to cut a cylinder-shaped tunnel into it. The opening inside an aerophone is called its *bore*. Start by slicing off the tip of the vegetable; make your slice about 1 inch (2–3 cm) long. The sliced end should be wider than the corer. Set it aside.
3. Twist the corer into the vegetable as far as you can without breaking through the other end. Be careful not to poke through the sides as you work.

FIGURE 1-57: Cut off the end of the vegetable and set it aside for later.

FIGURE 1-58: A carrot peeler with a corer tip can be used.

4. Remove the "core" you just cut. Since it is still connected at the far end, you will have to work it out carefully. One method is to poke a knife into the end, like sticking a screwdriver into the slot of a screw. Hold on to the knife while you try to twist the core until the end breaks off. If that doesn't work, slice through the center of the core all the way down and then work out each half. Use the knife or vegetable peeler to clean out any bits remaining and smooth the inside of the bore. Rinse the vegetable if it needs it.
5. Choose one side of the vegetable to be the top of the ocarina. Leaving enough room for your lips when you blow on the end, take the knife and slice straight down, but only until you reach the bore. Then make another cut at an angle a little way back from the first cut and toward it so that it connects with the bottom of the first cut, making a wedge. You should be able to remove this triangular-shaped piece to create a little window into the bore.

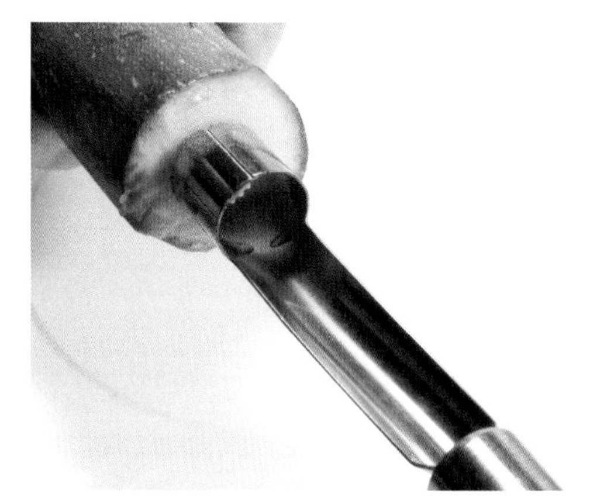

FIGURE 1-59: An apple corer used on a zucchini

FIGURE 1-60: Stick the knife into the core like a screwdriver and twist as you pull the core out.

FIGURE 1-61: Use a thin plastic knife to remove the pulp.

FIGURE 1-62: **Slice straight down to start cutting a window in the top.**

FIGURE 1-63: **Make a second cut at an angle.**

6. Next, you need to make a fipple—a plug that makes your instrument whistle. Take the leftover tip and trim it so it is roughly the width of the bore. If the vegetable is soft like a cucumber, you may be able to squish a wider piece into place. You want it to fit snugly and stay in place when you blow! The fipple should be long enough to reach from the end of the ocarina to the "window" at the top. Slice off the top of the fipple so it is flat. Then slide the fipple into the bore, with the narrower end pointing out and the flat side up.
7. Test your flute by blowing into the opening until you get a clear whistling sound. You may need to move the fipple in and out or higher or lower, or trim it a little more. Try different shapes and

FIGURE 1-64: **Remove the wedge for your window.**

FIGURE 1-65: **Trim and squeeze the end of the veggie until it fits inside the bore—but be careful not to tear your flute!**

placements. Don't get discouraged if your instrument doesn't work the first (or second, or third) time. You may need to make a few models to get one that makes a clear whistling note.

8. Once you have a good sound, you may want to carefully slice off an angled piece from the bottom of the opening to make it easier to fit the instrument in your mouth.

9. Add finger holes that will let you play different notes. On an ocarina, where the air churns around, it doesn't matter where the holes are. To make different notes, make them different sizes.

10. Try different vegetables to see what different sounds you get. You can even start your own band! But don't forget that your instruments are perishable. Just keep some extra veggies on hand so you can quickly carve up a replacement as needed!

FIGURE 1-66: **You can shave a slice off the bottom of the mouthpiece to fit.**

Bells and Beats 2

Create instruments that will make people want to get up and dance!

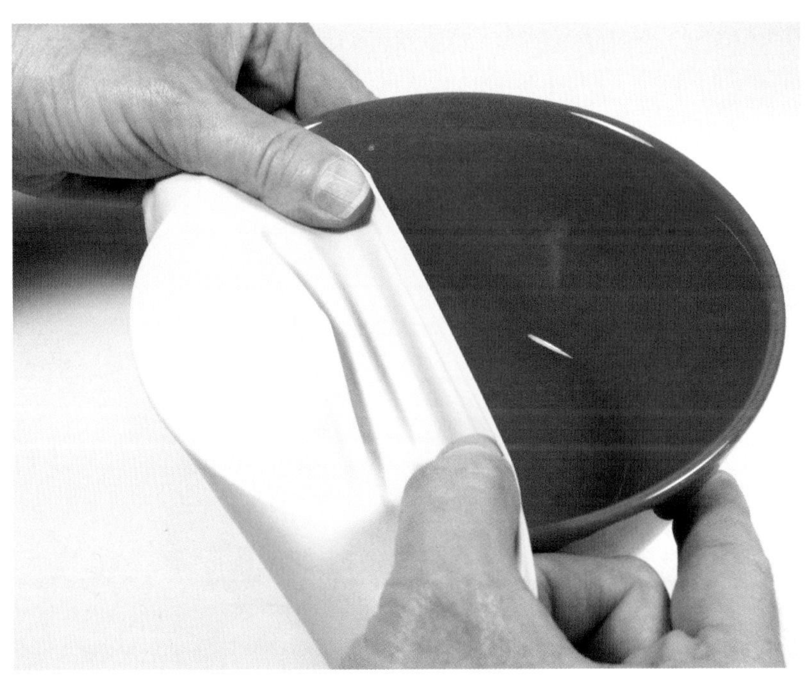

As you already know, music is more than just a stream of pretty sounds, such as the tinkling of chimes in the wind. And one of the things that makes it different is that music follows repeating patterns. The *melody* is the pattern of changes in pitch—how the notes go up and down. The string and wind instruments you made in Chapter 1 are great for playing melodies. But that is only part of what makes up a piece of music.

Music also has to have a repeating pattern that marks time. In most pieces of music, you can count the time in repeating groups of two, three, or four beats. The beat stays the same, no matter how long or short the notes are. The *rhythm* of a piece of music is the pattern of how the notes move along in time. Sometimes the rhythm moves along in step with the beat, but it can also make things interesting by making some beats sound stronger than others. Percussion instruments—drums and other types of instruments you play by banging on them—help keep the beat, but they are also important in creating the rhythm. The rhythm of a piece of music can be just as unique as the melody. If you can "name that tune" by hearing a few notes on the guitar or piano, you may also be able to recognize a song just by hearing the rhythm played on the drums.

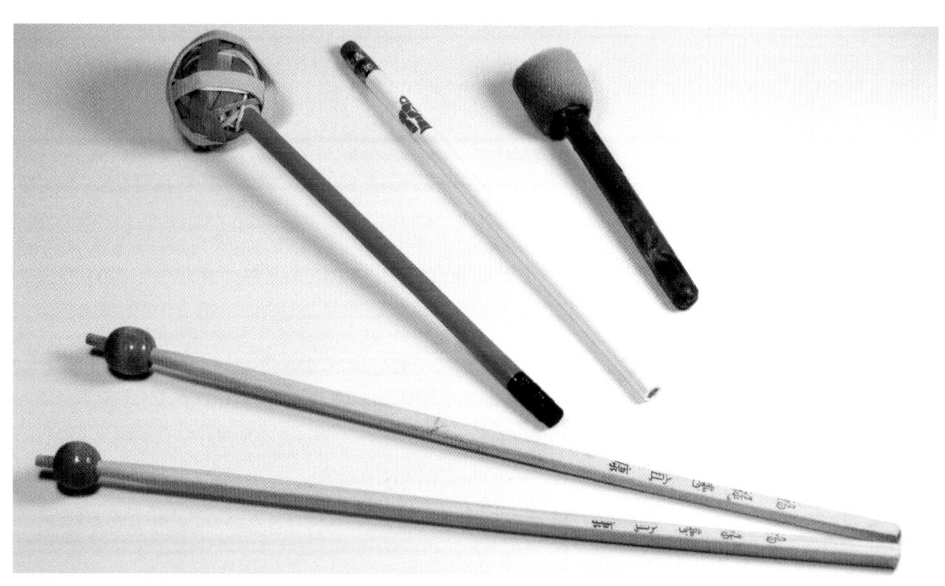

FIGURE 2-1: Different DIY mallets and sticks you can use to play your idiophones and membranophones. From left to right: chopsticks with wooden beads on the ends, a pencil with rubber-band ball on the end, a plain pencil with an eraser, and a foam paint brush.

Of course, percussion instruments do a lot more than thump. In fact, idiophones (instruments that vibrate throughout their whole body, like bells) and membranophones (instruments with a vibrating skin, like drums) may be the most versatile kind of instruments when it comes to changing the timbre of the sound they produce. One reason is that they can be made from almost any kind of material—wood, metal, leather, plastic, rubber, or cardboard. They can also be played in a variety of ways. You can tap them, rub them, or scrape them on their tops, bottoms, or sides, using your hands, sticks, or mallets. You can control how air affects their sound by making them closed or open, solid or hollow.

Some percussion instruments can even play a tune as sweet as any string or wind instrument. The pitch of an idiophone depends on its size and its mass, just like other kinds of instruments. On a membranophone, you can also change the pitch by changing the tension on its vibrating skin. There are even membranophones that let you change the pitch while you are playing.

Try making some of the idiophone and membranophone projects in this chapter and see how much sound you can really get out of them. Experiment with different materials and designs to find out how they affect the timbre and pitch. You may be surprised!

Got Rhythm?

There are a number of ways you can use rhythm to make a piece of music more interesting and even affect how it makes you feel.

You can make a song sound very different by singing it faster or slower than usual, or by changing the speed as you go along. The speed of a piece of music is called its *tempo*. Try singing a funny song at a slow crawl. (For example, the song that starts "If you're happy and you know it clap your hands.") Does it take on a heavy, serious tone? If you sing a sad song at a zippy tempo, does it make it more hopeful?

You can set the mood of a piece of music by choosing how many beats occur in each measure. (As discussed in the Introduction, a *measure* is a group of repeating beats.) The number of beats in a measure is known as its *meter*. In Western music, four beats in a measure is the most common. (An example is "Twinkle Twinkle Little Star.") When you have three beats in a measure, or 3/4 time, your music can seem to swoop and soar. This is the rhythm used for the dance called the waltz. ("Rock a Bye Baby" with its gently rolling rhythm is also in 3/4 time.) Two beats per measure, or 2/4 time, can sound like two feet stomping. That's the meter used for marches, which were originally written to help soldiers march together in formation. (Think of the song that starts "The ants go marching one by one. Hoorah! Hoorah!")

When you put *stress* on a note, you make it more important than the other notes around it. A waltz usually puts the stress on the first beat—ONE two three, ONE two three. In 4/4 time, the most stress is usually put on the first beat, but the third beat also gets a little stress. In a *syncopated rhythm*, the beats that are normally weak get the most stress. This kind of rhythm throws you off-balance and makes you want to move. It is often used in rock and jazz music.

Standing Waves, Overtones, and Harmonics in Idiophones and Membranophones

Idiophones include bells, cymbals, clappers like castanets, shakers like maracas, scrapers, and blocks. (The spoons and the washboard from the Jug Band described in Chapter 1 are idiophones.) Membranophones are mainly drums, although the kazoo (and the comb and tissue paper from the Jug Band) can also be considered membranophones. Just like strings and winds, these instruments follow the rule that bigger, heavier instruments (or bigger vibrating parts) produce lower notes. But the standing wave patterns they produce are usually much more complicated.

Instead of going back and forth like on a chordophone or aerophone, the sound waves created inside a hollow percussion instrument like a drum bounce off the round walls at all kinds of angles. As a result, many percussion instruments make noise rather than notes. The sound of a cymbal crashing or a bass drum pounding can be used in your musical piece, but they can't carry a tune.

Other kinds of idiophones and membranophones create layers of notes in the form of overtones. Think of the multiple notes you can hear when you bang on a gong (or a metal garbage can lid). The overtones sound fuzzy because they are not harmonics of the fundamental frequency. It's as if you hit a bunch of random keys on a piano keyboard at the same time. Instead of a chord of pleasant notes that points your brain toward the fundamental frequency, you get a jumble of unrelated notes that never quite blend.

But some kinds of percussion instruments produce distinct notes that are clear enough to create melodies. That's because their overtones are harmonic frequencies. In fact, to make them sound better, some instruments are specially designed to produce overtones that are closer to the harmonic.

Take the marimba, an instrument consisting of a row of wooden bars that looks very much like a xylophone. When you strike a marimba bar with a mallet, it creates a standing wave pattern that is related to the length of the bar. But the nodes (the parts of the standing wave that stay still) of the fundamental frequency are not on the ends, like they are on a guitar string. A guitar string is held in place at the ends. On a marimba, the bars rest on a support, and the ends are free to flap up and down as it vibrates. So the nodes for the fundamental frequency are further in toward the middle.

To make the note more pure, the bar on a marimba is thinned out underneath where the nodes for the fundamental frequency are found. The wood is more flexible at the thinner spots, so the bar tends to bend there. This bending action makes the fundamental frequency stronger and dampens other overtones, so the pitch sounds clear and true.

Note: To get the best sound out of your idiophones and membranophones, you need to let them vibrate. If you are using a drumstick or mallet, don't grip it tightly—let it dangle in your hand. If you are playing a drum, nestle it loosely under your arm or between your knees. If you want it to sit on a table or the floor, consider adding some feet or building it a stand to lift it up a bit and let the sound waves flow out. That way you'll get the maximum sound (and often a better type of sound, too!).

Project: Singing Bowl

Materials

Heavy metal bowl, no handles (Test it to make sure it rings for several seconds when tapped.)

Wooden mixing spoon (with as thick a handle as possible) or thin wooden rolling pin

FIGURE 2-2: **A Himalayan singing bowl decorated with an image of Buddha**

Did you know you can make metal bowls sing simply by rubbing them the right way? The singing bowl was originally used for religious ceremonies in Tibet, in the Himalaya mountains of Asia. Today it is also used in music for relaxation and meditation.

The typical singing bowl is made of bronze, a mixture of metals that can include copper, silver, nickel, and gold. It is usually small but very thick and heavy. To play the singing bowl, you rub a short, wide wooden mallet called a *puja* around and around the outside of the bowl. The puja is usually padded with a covering of leather or felt.

FIGURE 2-3: **Rubbing a cloth-covered puja against a singing bowl**

The singing bowl works thanks to *stick-slip motion.* As you rub the puja against the edge of the bowl, the molecules of the mallet stick to the molecules of the bowl for a moment—held together by the force of friction. Keep pushing and the mallet shoots forward just a hair, until it sticks again.

The sound of an authentic singing bowl is very pure. That's because there is only one standing wave, and it travels around the edge of the bowl, right behind the puja. As you play, you will notice that the tone goes up and down in volume. When it's softer, the node of the standing wave is passing by your ear.

For the best results, look for the heaviest metal bowl you can find. The stick you use as a puja should be as wide as you can comfortably hold, and very smooth.

1. Rest the metal bowl in the palm of your hand. Don't grip it; it has to be able to resonate. If the bowl is too big, you can set it on a table, but you may need to place a soft cloth or mat underneath it to keep it from sliding around.
2. Hold the wooden spoon with the flat part facing up. Grasp the end of the handle as if it were a very big pencil.
3. Tap the edge of the metal bowl with the side of the wooden spoon handle. Hit it just hard enough to make it ring.
4. To keep the vibrations going and help them build, start to rub the rim of the bowl with the side of the wooden spoon handle. Go all the way around in smooth circles, pressing the spoon against the bowl firmly. You don't have to go very fast. As the vibrations build, you will feel them in your hand. You should also hear a loud hum start to build. See if you can make the sound louder or softer by changing the speed of your mallet.

FIGURE 2-4: You can feel the vibrations in the metal bowl as you play.

Variations:

- Try using different-sized wooden mallets to see if the sound changes.
- Try rubbing the mallet on the lip of the bowl instead of the side to see if that changes the pitch. With a metal mixing bowl, you may be able to produce two or more different notes by changing the angle of the mallet.

Project: Tunable Water Glasses

Materials

Stemmed wine glasses (find them in thrift shops or garage sales)

Small wooden spoon or wooden pencil

Water

(Optional) paper or plastic plates

FIGURE 2-5: **Sound waves cause ripples in the water when you rub your finger around the edge of a glass.**

Musical glasses are usually thought of as a party trick, but in fact, famous classical composers like Mozart and Beethoven wrote pieces for glass instruments. There are two ways to play tunable water glasses, and this activity lets you try them both. The first way is to play them like a series of bells that you strike to get a tone. Different sizes and shapes of glasses will create different notes and different kinds of overtones. The other method uses just your finger and a little water to get the glass vibrating, just like the singing bowl in the previous project. Listen for the differences between the notes and the overtones that each style creates!

1. Line up the wine glasses in front of you in size order. Take the spoon or pencil and tap each glass to hear what note it makes. Larger glasses will usually produce lower notes—but not always! The thickness and shape of the glass may affect the notes as well. Rearrange any glasses so that the notes go up in order.

FIGURE 2-6: **Glasses will play different notes, depending on their size, weight, and how much water you put in them. These glasses are tuned to play the first seven notes in an octave!**

2. If you want to change any of the notes, add some water to those glasses. (To keep spills contained, you may want to place each glass on a paper or plastic plate.) Adding water lowers the pitch. Why? Because adding water is like increasing the mass of the glass—the glass and the water together figure into the speed of the sound wave's wavelength. You can tune the glass by adding or pouring out water until you get the note you're looking for. Depending on the glass, you may be able to produce several notes. (Bigger glasses seem to have a wider range.) As you fill the glass, you may notice that some notes sound better than others. This could be because the note is one of the harmonics for the fundamental frequency of the glass.

FIGURE 2-7: **Hold the "mallet" loosely and tap the glass lightly.**

MUSICAL INVENTORS: BENJAMIN FRANKLIN

In 1761, Benjamin Franklin—scientist, inventor, and one of the leaders of the American Revolution—took the idea of musical glasses and turned it into an instrument called the *armonica*. Instead of separate glasses sitting on a table, Franklin designed a series of glass bowls ranging from small to large. Each bowl had a hole in the center, and he attached them in size order to a rotating metal rod. The rod sat in a cabinet that was filled with water high enough to cover the bottom edge of the bowls. To play the armonica, you pumped a foot pedal up and down to turn gears that made the rod and the bowls spin slowly. As they turned, the water kept the edges of the bowls moist, so all you had to do was touch them to make them produce a note. Players could use all the fingers on both their hands to play different notes at the same time, just like pressing the keys on a piano keyboard.

The armonica was an overnight sensation in both the American colonies and in Europe, but its high-pitched vibrations led to its doom. Rumors started that the armonica could summon dead spirits, had magical powers, or drove listeners and players mad. It's possible that poisonous lead paint used to mark the bowls may have affected the players, if not the audience. In any case, the armonica disappeared from drawing rooms and concert stages as suddenly as it had arrived, and it became another footnote in the history of musical inventions.

3. When you have a nice selection of notes, try playing a tune with your glass organ.
4. Next, try playing the glasses by rubbing them. With one hand, hold the base of the glass. With the other hand, dip a finger in water. (If the glass has water in it, you can dip it right in the glass.) Rub your wet finger around the rim of the glass. As your finger moves around the rim, you will hear the glass start to emit a ringing tone. If

FIGURE 2-8: Hold the bottom of the stem to keep the glass steady, then wet a finger and rub it along the rim of the glass.

you have trouble getting started or keeping the tone going, re-wet your finger or adjust the pressure of your finger on the rim of the glass. You may have to press hard at first, but you should quickly figure out how much pressure you really need. It may not be a lot. (In larger glasses, you may be able to see ripples on the surface of the water as you play. This is a result of the sound waves traveling through the water.)

5. Try playing a tune again, this time using the rubbing method. Do you hear more overtones, or different notes, than when you tap the glass?

Project: Rainstick

Materials

- **Cardboard tube at least 1 inch (2.5 cm) wide, such as a wrapping paper or mailing tube (You can also tape two or more paper towel tubes together.)**
- **Pushpin**
- **60–90 thin nails or straight pins, slightly shorter than the width of the tube**
- **Masking tape (can be colored or patterned)**
- **Wax paper**
- **Decorative paper, ribbon, or tape**
- **Dried rice, beans, lentils, popcorn kernels, or other dried grains**

A *rainstick* is a South American instrument made from a dried cactus branch. The needles on the cactus are broken off and the points are pushed into the hollow branch. When the branch dries out, small pebbles are poured inside and the end is plugged.

As you tilt the rainstick, you hear a soft whooshing sound as the pebbles slide around and bounce off the needles. You can make a cardboard version and decorate it however you like. Make a few with different kinds of fillings, or combine smaller materials with larger materials to get different sound effects.

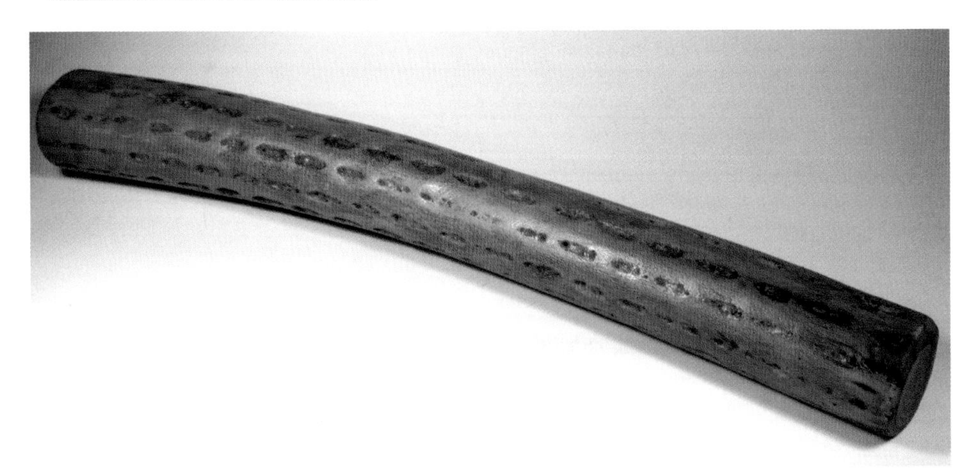

FIGURE 2-9: **A rainstick made from a dried cactus**

FIGURE 2-10: A rainstick made from a cardboard tube

1. Find the spiral seam that runs down the cardboard tube, or draw your own. Start at one end of the tube and use the pushpin to poke a line of holes along the seam. To make measuring easier, you can make the spacing between the holes the same width as one or two fingers. When you're done, insert a nail into every hole. (If the cardboard tube is very thick, mark the spots with a pencil, then use a hammer to pound the nails in.)
2. Keep the nails from falling out by sticking a strip of tape over the line of nails. If you like, cover the outside of the tube with decorative paper, ribbon, or tape.

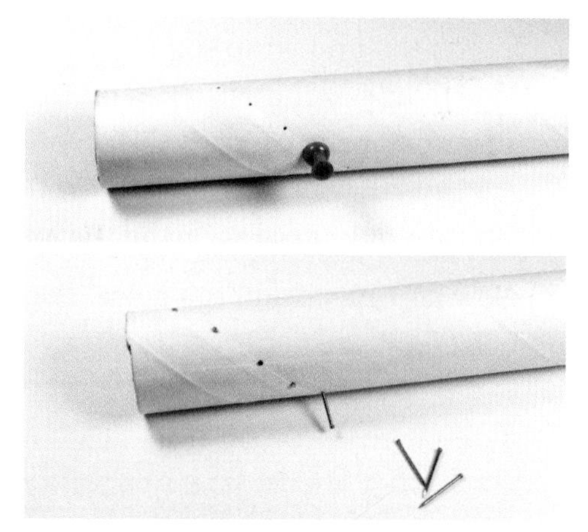

FIGURE 2-11: Make holes along the seam with a pushpin, then insert small thin nails into the holes.

FIGURE 2-12: Wrap tape over the nails to keep them in place. You can add more tape for decoration.

3. To make covers for the ends, fold a piece of wax paper in half. Place it over one end of the tube and press it down as tightly as possible. Remove the folded sheet and smooth out the wrinkles. Around the circle left by the tube, draw a bigger circle. Cut around the circle you drew to make two round end covers.

FIGURE 2-13: Crunch a double layer of wax paper around the end of the tube to make end covers.

4. Attach one of the end covers by folding it back down over the tube. Pull the paper as tightly as you can and wrap some tape tightly around it. Trim off the excess wax paper, and then use long strips of tape to connect the cover to the tube, going up the side, over the end, and down the other side. You want to muffle the sound at the ends a bit in addition to securing the cover.

5. Now it's time to add the noise. Pour in a handful of dried rice or other noise-making material. Test it by covering the open end with your hand

FIGURE 2-14: Press the wax paper circle over one end of the tube and secure it with tape.

FIGURE 2-15: Completely cover the wax paper over the end of the tube with more tape.

as you tilt the rain stick. If you need to, adjust the amount of rice until you get the sound you want. Then seal off the other end with the other round piece of wax paper. Tape it securely as you did for the first end.

Variation:

- Try out different types of dried grains as noisemakers. A combination of different sizes also sounds nice.

FIGURE 2-16: Try different kinds of dried seeds and beans as noisemakers.

Project: Cup Song

Materials

Sturdy plastic cup with a flat bottom

Smooth table (should be open underneath to help with resonance)

FIGURE 2-17: To get the cup song rhythm right takes a lot of practice.

The Cup Song project is all about rhythm. It's a variation of playground clapping games, but instead of using your hands, knees, and other parts of your body, you use a plastic cup and a table. As an instrument, the plastic cup is like the stamping stick, a primitive instrument used in many ancient cultures. The stamping stick is played just as you would expect, by taking a large stick and banging the end on the ground to create a thumping sound. But because the cup is hollow and lightweight, you can create a range of different sounds by banging it on a table, against your body, and even just sliding it or running your hand along it.

The cup song became a viral hit when actress Anna Kendrick used the cup song rhythm to keep the beat as she sang the song "You're Gonna Miss Me" in the 2012 movie *Pitch Perfect*. She learned it by watching a YouTube video inspired by the group Lulu and the Lampshades (now called Land-Shapes). (Watch the video here: *https://youtu.be/DWCOYJg9ps4*.) The tune was originally recorded in 1937 by A. P. Carter under the title "When I'm Gone," and it may have been based on an even older traditional folksong.

Try to master the popular cup song rhythm in the following activity, and then see what new rhythms and songs you can create with musical cups!

1. The cup song rhythm consists of a series of moves, some of which are repeated in different ways. They include clapping your hands together, tapping them on the bottom of an upside down cup, banging the rim of an upside down cup on the table, bonking the rim of the cup against your open palm, and banging a "corner" of the bottom of the cup on the table. You can even use sounds produced by your hands sliding on the cup as you pass it from hand to hand. Here are the moves of the first part (to get the rhythm, see the sidebar that follows this activity):

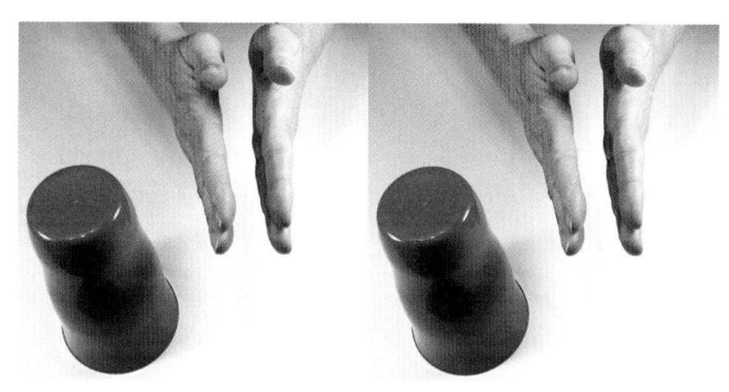

FIGURE 2-18: Clap, clap.

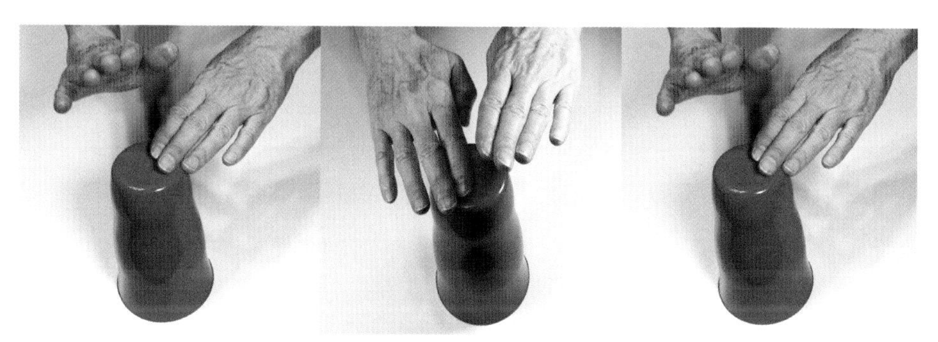

FIGURE 2-19: Tap, tap, tap.

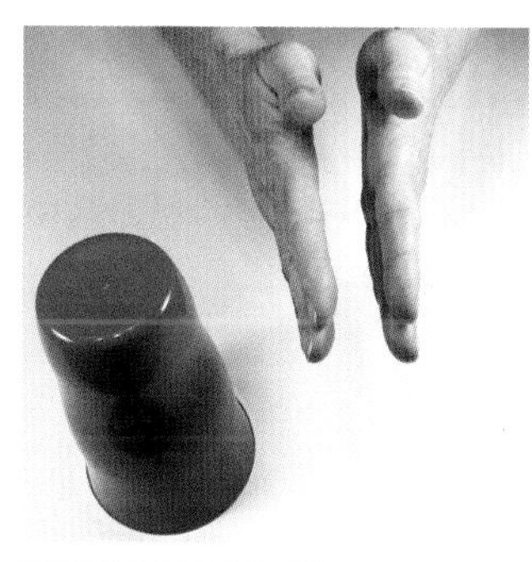

FIGURE 2-20: Clap.

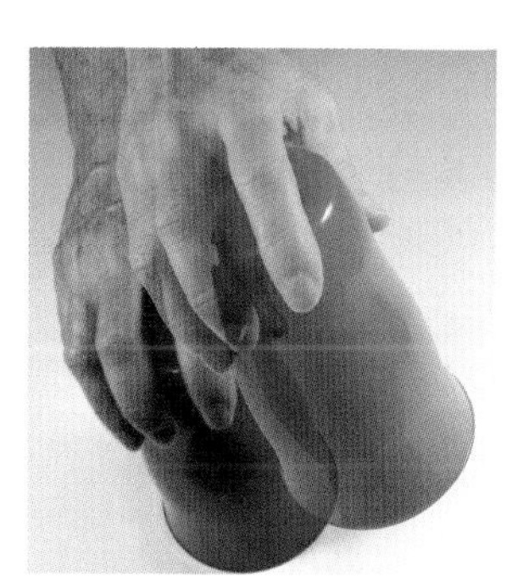

FIGURE 2-21: Pick up the cup, move it to the side, and bang the rim of the cup flat against the table.

2. For the second part, you have to lift the cup with the same hand as before, but make sure your thumb is pointed down. These are the next moves:

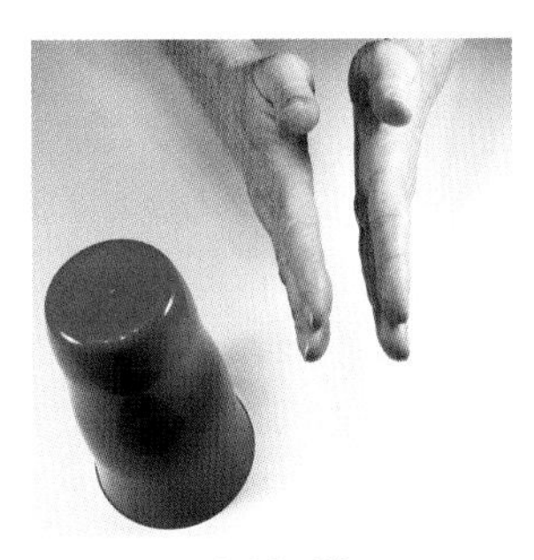

FIGURE 2-22: Clap.

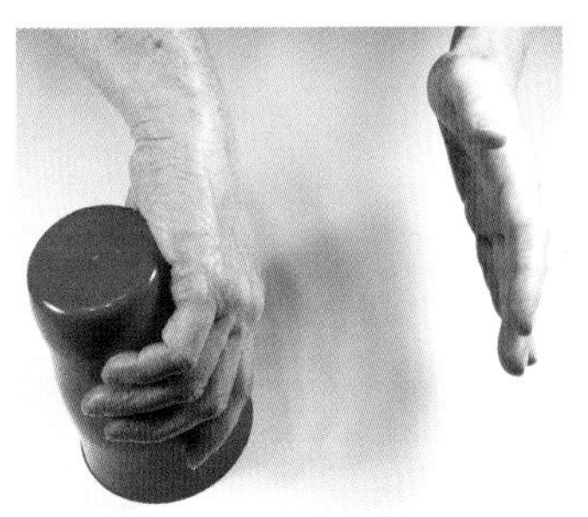

FIGURE 2-23: Lift the cup with your thumb pointed down.

FIGURE 2-24: Bop the opening against your palm.

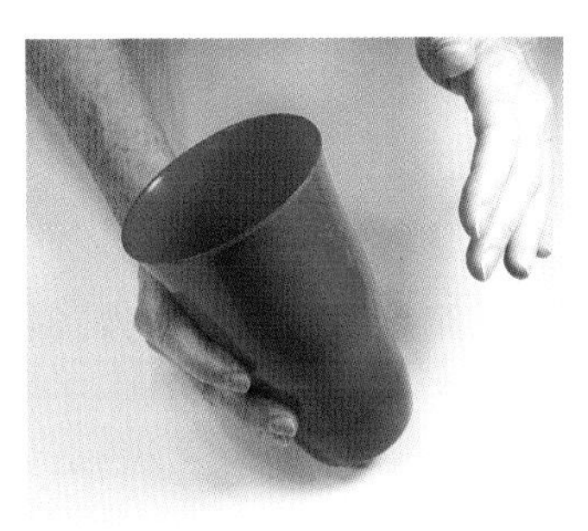

FIGURE 2-25: Knock the bottom "corner" of the cup against the table.

FIGURE 2-26: Switch the cup from one hand to the other—crossing your hands as you do so! (This step doesn't need to make a sound, but it must be done quickly and in rhythm.)

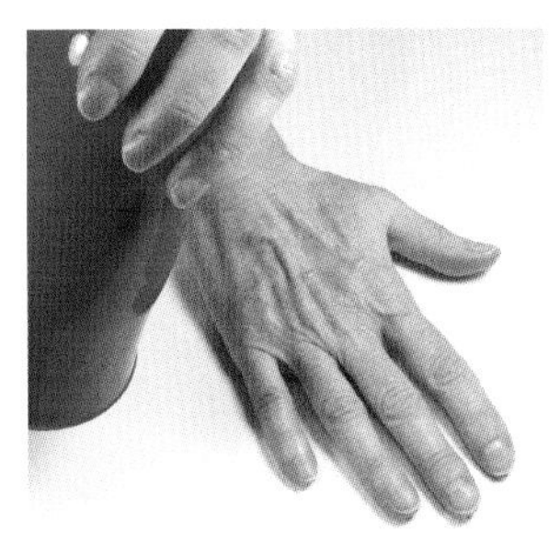

FIGURE 2-27: Smack your free hand—which crosses in front of you at the moment—on the table.

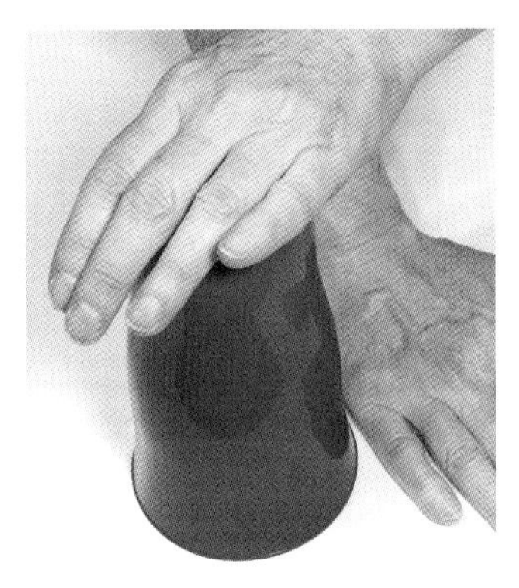

FIGURE 2-28: Bang the rim of the cup flat against the table.

CUP SONG RHYTHM

The rhythm of a drumbeat can be written with notes on a staff, just like other kinds of music. But an easier way to learn and remember the rhythm of a clapping game is to make up a sentence that has the same pattern of beats. Here's how you might translate the rhythm of the cup song into words:

Glass of lemonade with a straw.

Please bring it quick, don't wait!

Here's how the words match the motions:

Glass of le- mon- ade with a straw.

(clap) (clap) (tap)(tap)(tap) (clap) (pick up) (bang)

Please bring it quick, don't wait!

(clap) (pick up) (fwap) (knock) (silent switch) (smack) (bang)

3. The cup should now be back in the starting position. Repeat the rest of the steps in order as many times as needed for the song. (Again, see the sidebar for help with the cup song rhythm.)
4. *Adaptation:* If the moves are too hard, try lifting and banging the cup on the same spot (don't move it to the side). When you pass the cup to your other hand, don't cross them—just slap the table in front of your free hand.

Extension and variations:

- Try to sing "You're Going to Miss Me" or other folksongs with the cup song rhythm. Here are some suggestions:
  - "Red River Valley"
  - "You Are My Sunshine"
  - "Shoo Fly, Don't Bother Me"
  - "Down By the Riverside"
  - "This Little Light of Mine"
  - "This Land Is Your Land"

♫ Ready for a cup song challenge? Get a few friends and sit in a circle or around a small table where everyone can reach the person next to them easily. When you pass the cup from one hand to the next, grab the cup from the person next to you and give yours to the person on your other side—all while singing your favorite cup song.

Project: Cookie Tin Steel Drum

Materials

Pencil with eraser (or with a slip-on pencil cap eraser)

Round cookie tin, 6 inches (15 cm) or wider

Permanent marker

Tools

Hammer (regular or jewelry-making ball-peen with a rounded hitting end)

FIGURE 2-29: **Steel drums started as recycled instruments but became accepted for their unique sound.**

Steel drums, also known as steel pans, come from the Caribbean island of Trinidad. They were invented in the 1800s, when the British government, which then ruled the island, banned real drums because they were used by criminal gangs. Instead, steel barrels used for storing oil were turned into makeshift instruments and used in local parades. In the early 1900s, players discovered they could create different notes by making dents in the top of the barrels, and it was then that the steel drum was born.

The simplest steel drum you can make is a dudup. It has only two notes and is used as a rhythm instrument. Try making a dudup before you move up to a drum with more notes. This project was adapted from directions on *www.toucans.net*.

Safety Warning: Put a piece of scrap board under your tin before you hammer it to avoid damaging a tabletop or wood floor. As always, watch your fingers when you use the hammer, and make sure children have adult supervision.

1. To make the steel drum, remove the lid from the cookie tin (save it for later so you can store your mallet inside the tin) and turn the cookie tin upside down. With the marker, draw a line across the bottom of the cookie tin that divides it into two parts, one a little larger than the other.
2. With the hammer, carefully tap along the line to stretch the metal downward. You should end up with a shallow valley running along the line.
3. Time to test the drum. Balance the tin on the palm of your hand (if it's too large, lay it across your lap, or set it up on two piles of books with a space in the middle). Hold the mallet (a pencil or other DIY drum stick) loosely in your other hand. Then tap on one side of the line and then the other to see what pitches you get.
4. To adjust the quality of the sound as well as the pitch, you can hammer the line down a little more. You can also hammer each side separately from inside the tin. This will create a little "hill" that will stretch the metal even more. As you know from working with string instruments, more tension

FIGURE 2-30: Make the line a little off-center so one side is larger than the other. Each side will produce a different note.

FIGURE 2-31: Hammer along the line to divide the surface into two notes.

FIGURE 2-32: Playing the steel drum

means faster sound waves. Your dudup is ready when it produces two clearly different notes on each side. Ideally, one side should be about four or five notes lower than the other; however, the exact pitch doesn't matter, because the dudup's job is just to create a beat with its two notes.

Extensions:

- Divide your drum surface into three or more sections to create extra notes.
- Build a stand for your drum. The string loops just described can also be used to hang the drum on a stand with arms on either side. Make a box-shaped frame for your drum from PVC pipe or wood. Add hooks to let your drum hang.
- Try some different mallets. Real steel drum mallets have relatively short handles that let you tap the drum at a close angle. They also have a smooth or round rubber tip. Try a small foam paint brush, or take a short dowel and attach a rubber ball or make a rubber-band ball on the end.

Project: Wrenchophone

Materials

Metal wrenches in a range of sizes, as many as possible (an inexpensive set of 18 costs about $10)

(Optional) other metal objects that can be used as bars, such as large bolts, dinner knives, and so on

6 chopsticks or square dowels

Rubber pencil grippers or other foam padding

2 large wooden beads

(Optional) white glue, rubber bands, zip ties

The Wrenchophone is a metallophone, a xylophone-like instrument with metal bars. (A true xylophone has bars made of wood.) Surprisingly, the metal tools that make up the Wrenchophone produce beautiful ringing tones. That's because you are hearing mainly the fundamental frequency, without other overtones that make the timbre fuzzier. The base you put together to hold the bars helps to dampen the other overtones that might get in the way of the fundamental frequency. Try to make this Wrenchophone and see what kind of tones you can produce using found objects!

Note: This instrument is temporary, so feel free to borrow wrenches and other tools from your tool chest. You can go back to using them for their original purpose when you're done playing music on them!

1. Take the wrenches or other metal objects you are using for your bars and lay them in a row, longest to shortest. Space them so they are easy to play. This will give you an idea of the size and shape of your Wrenchophone.

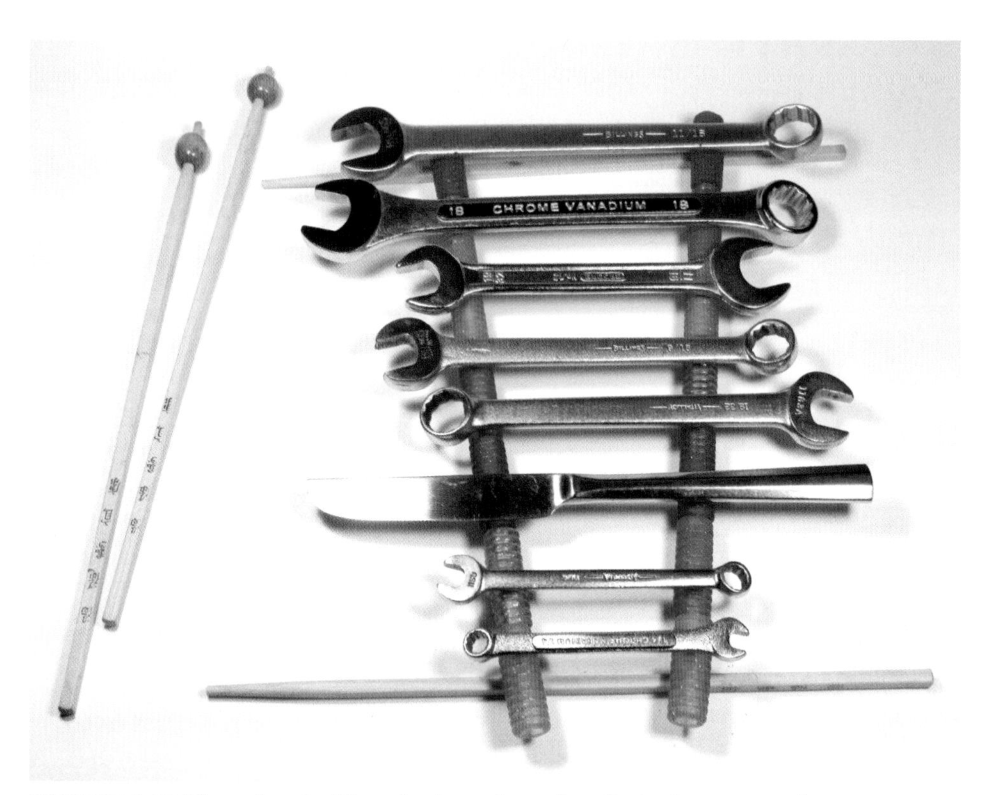

FIGURE 2-33: The notes of a Wrenchophone depend on the tools you use as bars.

2. Can you figure out the best place to support the bars? That's right, underneath the nodes, the parts of the standing wave that do not move! To find the nodes, measure each bar, and put marks at the spots that divide each bar into quarters. (You can do this quickly by eye by dividing the bar in half and then dividing each half in half.) The nodes for the fundamental frequency—the main note you hear when you set the bar vibrating—are near the outside marks, at the either end of the bar. (To be exact, the marks are 22 percent of the length in from each end of the bar. That is a little less than 25 percent, or one quarter, of the length.) For example, if your wrench is 12 inches (30 cm) long, the supports should go at about 3 inches (7.5 cm) and 9 inches (22.5 cm). If it's 4 inches (10 cm) long, they should go at around 1 inch (2.5 cm) and 3 inches (7.5 cm).

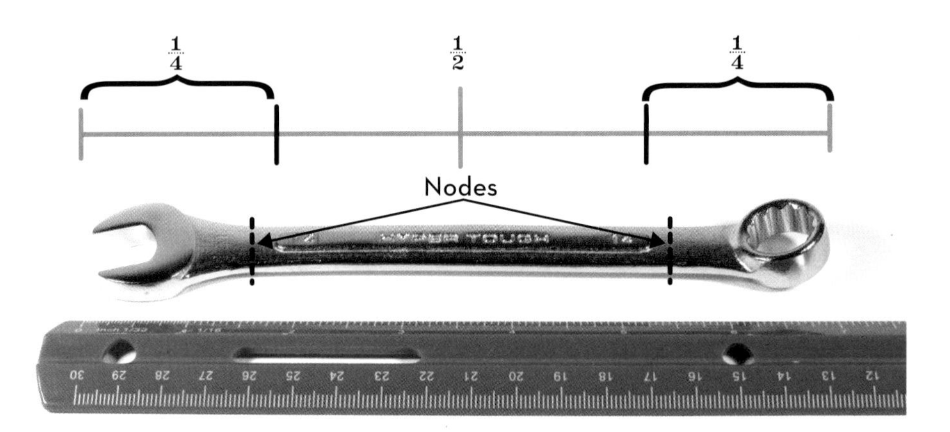

FIGURE 2-34: The nodes of the fundamental frequency of a bar on a xylophone-type instrument are found a little less than $\frac{1}{4}$ of the way from each end.

3. The bars of your Wrenchophone need to rest on a base that will support them but also let them vibrate. Unlike string instruments, the ends of the bars are loose and also vibrate up and down. So your base has to let both the middle and the ends move up and down. You can make a quick and temporary base by laying two chopsticks horizontally (side to side), and resting two more chopsticks vertically (up and down) on top of them. To cushion the sticks on top, slide some rubber pencil grippers over them, or wrap them in foam padding. The cushioning helps dampen the overtones from interfering with the fundamental tone of the metal bars. Place the bars on the cushioned chopsticks so that the marks sit above the supports. As the wrenches get smaller, the chopsticks will need to get closer, until they are almost in a V shape.

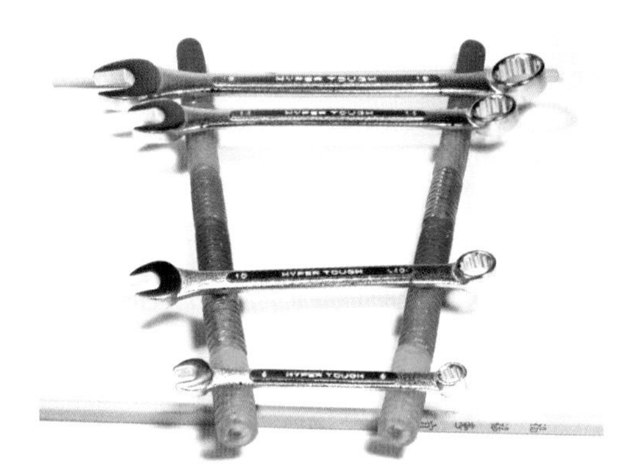

FIGURE 2-35: As the bars get shorter, the distance between the supports gets smaller.

4. Make a pair of mallets by sliding a large wooden bead onto the narrow end of each remaining chopstick until it is tight. Use a little white glue to hold it in place if you need to.

5. Test your Wrenchophone to see how it sounds. If you can, try to find bars that will create at least one full octave. This will allow you to play many songs. However, if you can't manage to get every note, pick a nice arrangement of notes that sounds good to you. You may need to add or subtract bars, or rearrange them in a different order. If you are mixing wrenches made by different companies, you may find that the variation in weight between them means that a shorter wrench plays a lower note than a longer one.

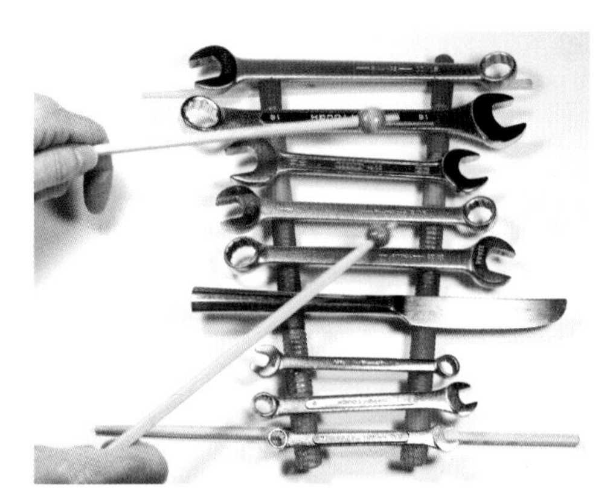

FIGURE 2-36: Use wooden mallets made from beads or other kinds of sticks to test your Wrenchophone.

6. Once you have the chopsticks where you want them, you can connect them permanently if you like with rubber bands, zip ties, or glue at the places where two chopsticks cross.

Variations:

- Try using different mallets to see what kind of sounds you get, such as the pencil or rubber-band-ball mallets from the Cookie Tin Steel Drum project. You can even just use a leftover wrench as a mallet.
- Try making a xylophone-type instrument out of other materials. Remember, if you are making your own bars and they are all made of the same kind of material, you can figure out the harmonics by measuring out simple fractions ($\frac{1}{2}$, $\frac{2}{3}$, $\frac{3}{4}$) of the longest bar. Here are some suggestions for materials to try:
  - PVC pipe
  - Copper pipe
  - Tree branches or driftwood

- Bamboo poles
- Stones (use a rock as a mallet)
- Pieces of ceramic tile

- Try other ideas for supporting the bars. Here are a couple of suggestions:
  - Suspend the bars from strings; if you drill holes through the bars, place the holes at the nodes.
  - Place a line of cup hooks along two strips of wood and hang rubber-band "hammocks" between them to support the bars.

Project: Packing Tape Bass Drum

Materials

Round can, stiff cardboard container, or thick cardboard tube, at least 6 inches (15 cm) wide

Clear packing tape

(Optional) decorative paper or tape, such as colored or patterned duct tape

Tools

Can opener

FIGURE 2-37: A packing tape drum sounds best when the body is big and heavy.

Making a membranophone requires a material that can be stretched tight but stay stiff enough to make a booming sound when you tap on it. The earliest drumheads were made of animal skins. Today many drumheads are made of sheets of plastic. The drumhead on this simple bass drum is made of clear plastic tape that is stretched across the openings. Hitting one of the skins causes the air inside to vibrate and makes the skin on the other side vibrate too. This helps amplify the sound, making the bass drum one of the loudest kinds of (nonelectric) instruments. This project was adapted from a YouTube video by Alec Duncan of Child's Play Music.

1. If the can or container you are using has a metal bottom, use the can opener to remove it. Your drum needs to be open at the bottom.
2. Choose one end to be the top. You are going to start to make the drumhead by stretching a piece of packing tape straight across the opening so that the tape goes through the center of the circle. Attach one end of the tape a few inches (5–10 cm) below the edge. Stretch the tape as tightly

as you can as you bring it across to the other side of the can. Then press the same amount of tape down over the side before cutting it.

3. Repeat with a second strip of tape, but turn the can so the new strip forms an X with the first piece. Add a third strip between the first two strips. Then add a fourth strip in the remaining gap between strips to make another X.

FIGURE 2-38: Stretch a piece of packing tape across the top of the drum.

4. Continue layering strips of tape across the opening the same way until the entire top of the can is covered. Be careful not to leave any openings. Then go around and make a second layer all the way around. When you have two complete layers of tape across the top of the can, turn it upside down. Press on the tape from the inside (sticky) side to make sure all the layers are stuck together tightly.

FIGURE 2-39: Make an X with a second piece of tape.

FIGURE 2-40: When finished, your drum should have at least two layers of tape across the top. Press them together from the inside.

5. Use a mallet (see the "Cookie Tin Steel Drum" or "Wrenchophone" projects) or your hand to play a rhythm on your drum. See what kind of sounds and beats you can get out of it!

Extension:

- ♫ Try making several drums using different containers in different sizes.

FIGURE 2-41: Dress up your cardboard drum with duct tape or other coverings.

Project: Balloon Drum

Materials

- Tin can, ceramic or wood bowl, or other study container
- Latex balloon, large enough to stretch over the mouth of the container (12-inch [30-cm] and 18-inch [45-cm] balloons can be found with party supplies)
- (Optional) balloon pump
- Electrical tape or large rubber band
- Pencils, chopsticks, or drum mallet (see directions for making these in other projects in this chapter)

FIGURE 2-42: **Any closed container, bowl, or mug can be used to make a balloon drum.**

For this project, you will create a kettledrum-type instrument. Instead of a tube (either open or with drumheads covering openings on both sides), kettledrums (also called tympani) consist of one drumhead that seals the opening of a bowl-shaped body. Sound waves bounce around inside, smoothing out the tone. In fact, real kettledrums are tunable. Players use winding keys like the pegs on a string instrument to make the drumhead tighter or looser, which raises and lowers the pitch. Kettledrums even have pedals that allow them to play multiple notes. See what kind of notes you can get with a stretchy latex balloon drumhead that can be pulled tighter over the opening to change the pitch.

Note: To avoid tearing the drumhead, make sure the drumsticks or mallets you use don't have any splinters or sharp edges!

1. To make the balloon easier to stretch, blow it up first and let the air out. Then cut the neck off the balloon.
2. Start to stretch the balloon over the opening of your can. It may help to hold the can between your knees and to have another person on hand to grab the balloon from the other end. "Hook" the edge of the balloon over the rim of the can closest to you. Then pull the two opposite "corners" until you can stretch them over the rim of the opening on the far side. If you have trouble getting the balloon on, or making it stay on, you may need a bigger balloon (or a smaller can or bowl).
3. Test the sound with your hand, a stick, or a mallet. To tune your drum, you can carefully pull the edge of the balloon up or down to make it fit tighter or looser around the can.
4. Once you've got it adjusted to your liking, wrap a rubber band or strip of electrical tape around the edge of the balloon to hold it on. If your drum body is a round bowl, the balloon may stay on by itself.

FIGURE 2-43: **Cut the balloon where the neck starts to widen out.**

FIGURE 2-44: **Carefully stretch the balloon over the mouth of the can or bowl.**

FIGURE 2-45: **If you have trouble, get another person to help you stretch the balloon over the opening.**

Variations:

- Use plastic food wrap instead of a balloon for your drumhead. Or, for a really tight fit, use plastic shrink wrap that's made for insulating windows. You'll have to tape it on first, then use a hair dryer to shrink it until it makes a good sound.
- Try making a set of drums using different containers in different sizes.

FIGURE 2-46: You can make temporary drums using dishware.

Mechanical Music

3

Build machines that make music—including some that play themselves!

When you start to invent your own instruments from everyday objects, your brain begins to pay extra attention to ordinary sounds. If you've ever heard the *boing* of a door stopper spring when you hit it with your shoe, the *scritch scritch* of an old pair of scissors, or the *knock knock knock* of a window shade in the breeze and thought "That sounds like the beginning of a song!" then you'll enjoy the projects in this chapter.

Some of the musical machines you'll learn about in this chapter are based on popular contraptions from the past. Others are inspired by artists of today who have come up with unique ways to turn accidental sounds into music. You'll need to use concepts like standing waves and nodes that you encountered in earlier chapters and combine them with elements of simple levers, springs, and more to make your musical machines move and play. You'll also get to think about what new sounds you can create and unusual ways to make them happen.

Shake Things Up with Resonance and Timbre

As you know, instruments create music by vibrating, which causes the air around them to vibrate. To make a soft instrument sound louder, it helps to add a resonator. A resonator vibrates at the same frequency as the note you are playing, which adds energy to the sound wave and raises the volume. If an instrument doesn't have a resonator built in, you can use another object as a resonator to help increase the sound.

You can use this trick with several of the projects in this chapter. For example, the thumb piano is a quiet instrument that you hold in your hand and pluck. Place it on a box, inside a bowl, or even on a table, however, and suddenly the sound explodes. In Africa, where the thumb piano got its start, players use hollow gourds to amplify the sound. If the inventions you create are hard to hear, try setting them on top of an empty box or leaning them against a piece of furniture and see if the sound improves.

Something else you will discover as you look for objects to add to your inventions in this chapter is how different materials can affect the tone. The timbre of an instrument is the type of sound it makes. As you experiment with various items, pay attention to the way metal, wood, glass, and other materials make your instrument sound—even when the notes are the same. You may even want to make several versions and compare them to see how they give your instrument its own personality.

Project: Thumb Piano

Materials

A body, for example:

- Block of wood, about 6 inches (15 cm) long and narrow enough to hold comfortably in your cupped hands, such as a piece of 1×3 (2×6 cm) lumber
- Small metal candy tin
- Wooden box

5 or more tines; for example:

- Large bobby pins (hair pins made of thin flat wire)
- Wooden coffee stirrers
- Craft sticks
- Thin bamboo skewers
- Chopsticks (the pointed end)
- Bicycle wheel spokes

Rods to hold the tines (2 for bobby pins, 4 for other tines), slightly longer than the width of the block of wood, and strong enough to resist bending. Possibilities include:

- Pencil stubs
- Chopsticks, cut to size with wire cutters or a utility knife (use the square end of the chopstick)
- Door hinge pins
- Large bamboo skewers
- Craft sticks or tongue depressors

Connectors, such as:

- Rubber bands
- Zip ties
- Pushpins
- Heavy staples

(Optional) a resonator, such as:

- Wooden box (craft stores sell unpainted versions)
- Metal or ceramic bowl
- Large cardboard canister
- Clay flowerpot

Tools

- (Optional) sandpaper
- (Optional) saw or utility knife
- (Optional) wire cutters or scissors

FIGURE 3-1: An empty box is used as a resonator for a thumb piano made from a solid piece of wood.

The thumb piano is a traditional instrument from Africa, where it is known by several names, including *kalimba* and *mbira*. It is a type of *lamellophone*, an instrument with strips or rods that you pluck. The thumb piano consists of a small body with a row of tines on the front of the instrument that are attached near the top. These are similar to the keys on a regular piano—each one plays a different note. To play the thumb piano, you hold it in both hands in front of you, like the way you hold your cell phone when you are texting, and press down on the tines with your thumbs. As the tines spring back up, they vibrate from the point where they are attached (or the bridge, if there is one) down to the end. The pitch depends on the length of this vibrating segment. Unlike most instruments, the notes on a thumb piano do not usually go from lowest to highest. Instead, they are arranged in a V shape or spread out like a folded paper fan. The longest tines that play the lowest notes are in the middle, and the shorter tines that play higher notes are on the sides.

The following directions show you how to make a thumb piano that doesn't require special tools and a kind of bridge that lets you use a variety of common materials for the tines.

Safety Warning: Children should get adult help to cut materials like chopsticks or metal strips. Be extra careful when cutting the tines, as pieces can fly off in any direction. Eye protection is recommended. Use sandpaper to smooth off rough edges on cut materials.

1. First, choose a body. It should be small enough to hold in your hands easily, about the size of a cell phone (but thicker). When deciding on the width, also think about how many tines (which you will choose in the next step) you can fit in a row across the body.
2. The tines you choose for your thumb piano should be narrow enough to allow you to fit five or more across the width of the body. They can be longer than the body if you don't mind having the excess hang off the top. If you want to cut them shorter, wire cutters work well for most kinds of materials. (See the earlier Safety Warning.)

FIGURE 3-2: Try different kinds of bendy material for tines. This thumb piano was made with old bicycle wheel spokes. The rods are door hinge pins.

3. Choose rods that will hold your tines down securely, such as pencil stubs or chopstick pieces. The secret to making your thumb piano sound good is to attach the tines as tightly as possible! The rods should be a little longer than the body is wide. For the bobby pin method, you need two. For other materials, you need four.

FIGURE 3-3: **The bobby pins in the middle are lower than those on the ends. Insert a rod inside the pins. Place another rod underneath and connect them loosely with rubber bands.**

4. Time to attach the tines! If you are using whole bobby pins, the method is simple: place the bobby pins in a row near the top of the body with the flat side down. Arrange the pins so their ends make a V shape. Insert one rod inside the pins so it lies across the entire row. (*Important: Do not try to bend the pins open—let the rod push them open.*) Place the other rod underneath the body so its ends line up with the top rod. Wrap a rubber band around one end of the top and bottom rods. It should be just tight enough to hold the rods in place. Repeat with the other end. At this point you can try plucking the tines to get an idea of what pitch they play. You can adjust them now, but you will tune them exactly later. To finish, take another rubber band and wrap it around the end of the rods on each side, right over the first rubber band. This time make it as tight as possible. The two ends should be equally tight.

FIGURE 3-4: **The rods on this thumb piano are made from pieces of chopsticks. When you are done tuning the bobby pins, tighten the rods with more rubber bands.**

5. The method for other kinds of tines is a little different. Start by connecting the rods above and below the body with the first set of rubber bands. Then slide the tines under the top rod. Next, take two more rods

and slide them under the tines. They should go on either side of the rods with the rubber bands. The rods act like the bridge on a guitar to hold the tines away from the body. Finish by tightening up the top and bottom rods that are holding the tines in place with the second set of rubber bands.

FIGURE 3-5: **For most tines besides bobby pins, attach the rods first.**

6. Now you can tune your thumb piano. Lean the instrument against a table, or set it on top of a hollow box or other resonator (see suggestions in the Materials list). Hold it in both hands and use your thumbs to pluck each tine by pressing it down briefly. Adjust the tuning of each tine until you are happy with all the notes. To raise the pitch on a tine, push it toward the top of the thumb piano. This shortens the part of the tine that is vibrating. To lower the tone, push the tine toward the bottom of the thumb piano to make the vibrating section longer.

FIGURE 3-6: **Slide extra rods under the tines to hold them away from the body.**

Troubleshooting tips:

- The tighter the tines are held down, the clearer the vibrations. If none of the tines sound good, try squeezing the top and bottom rods closer together. If you used rubber bands, loop

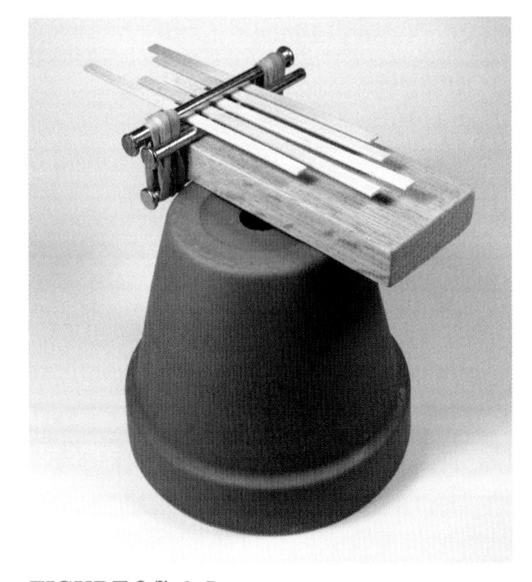

FIGURE 3-7: **A flowerpot resonator**

them around one more time. Or use a zip tie to pull them tighter.

- If just one or two tines are loose, look for replacement tines that are a little thicker.
- If a tine on the bobby pin version is buzzing, check to see if the upper end of the bobby pin (where it bends) is raised off the body. You can fasten it down so it is held tight against the body by looping a separate rubber band through the bobby pin and around the board. Or run a craft stick or coffee stirrer through all the bobby pins near the bends. Attach it to another stick under the body, as you did with the rods.

FIGURE 3-8: The craft sticks near the top of the bobby pins are there to hold them tighter to prevent buzzing.

- If you can't get one particular note to sound good no matter what you do, it may be causing sympathetic vibrations in another part of the instrument that muddy the sound. Try skipping that note altogether and make that tine higher or lower in pitch. Shift it around until you find a position that sounds better.

Project: Musical Marble Run

Materials

1 or more marbles or steel ball bearings, about ½ inch (1.25 cm) in diameter

35 or more bamboo barbeque skewers, 12 inches (30 cm) long

2 wooden tongue depressors (large craft sticks)

String (elastic string works well)

(Optional) small rubber bands (if your string isn't stretchy enough)

Assorted "musical" objects, such as these:

- Bells
- Mini wind chimes
- Old keys
- Metal washers
- Wooden beads
- Wooden craft sticks of various widths
- Bottle caps

Tools

Hot glue gun

Heavy-duty wire cutters, scissors, or craft knife

Pencil

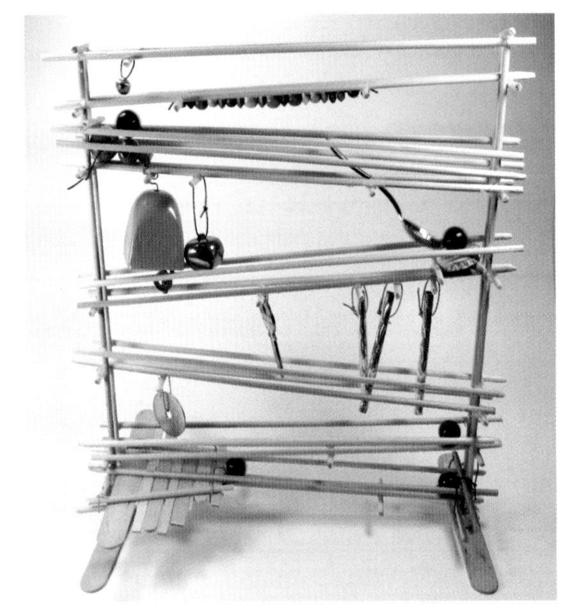

FIGURE 3-9: See how many different kinds of sounds you can create just by letting marbles roll and drop!

Marble runs are fun to watch—and when you add your own bells and shakers, they're also fun to listen to! This barebones marble run may seem simple, but it's got what robotics scientists call a "programmable body." That means you control the sounds it plays by how you place the ramps and noisemakers. Use the power of gravity and the concepts you learned about how instruments such as idiophones work to make it play faster or slower, softer or louder, and higher and lower. That's how you turn a basic toy into a real Musical Invention!

MUSICAL INVENTORS: WINTERGATAN

Martin Molin of the Swedish band Wintergatan (*www.wintergatan.net*) is known for the gigantic hand-powered marble music machine he designed and built out of wood, LEGO pieces, plastic tubes, funnels, rice, and drink coasters. When Molin cranks the machine up, 2,000 marbles are pulled to the top on little elevators. He programs the machine using belts made of LEGO Technic pieces. The pins open gates that allow marbles to drop onto the machine's built-in vibraphone (a xylophone with a wavering sound), drums, cymbal, or bass guitar in time with the music. Wintergatan's wildly popular music video featuring the Marble Machine has racked up more than 40 million views on YouTube (*youtu.be/IvUU8joBb1Q*).

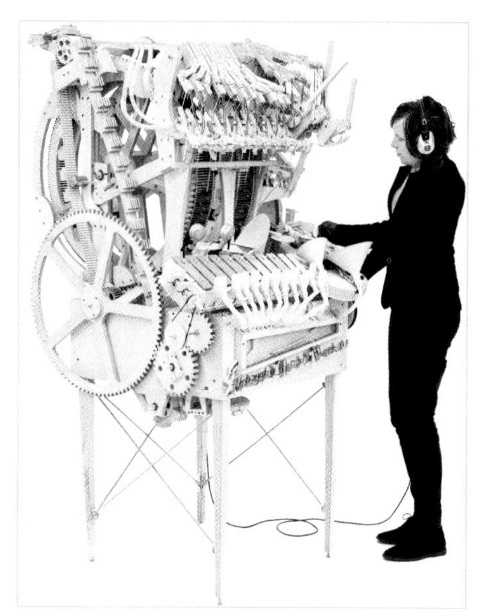

FIGURE 3-10: The hand-cranked giant musical marble machine built by Martin Molin of Wintergatan became a hit on YouTube in 2016. Photo credit: Samuel Westergren, courtesy of Wintergatan

Safety Warning:

- Be careful when cutting the bamboo skewers—sharp bits of wood can go flying off in unexpected directions. Protect your eyes by wearing safety glasses.
- Children who are not experienced with hot glue guns should get adult assistance. The heated tips and the melted glue can cause minor burns. Always use a piece of heavy scrap paper or a silicone pad under the hot glue gun to avoid damaging your work surface.

1. The bamboo skewers you are using for your building material have sharp points. Use the wire cutters to snip the points off. Wear eye protection and point the skewer point away from yourself and other people before you snip.

2. Also cut several skewers into pieces 1½ inches (4 cm) long. Set these aside where you can grab them as you need them.

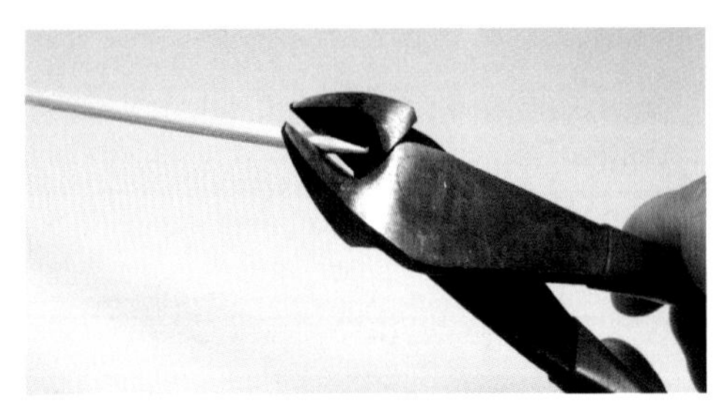

FIGURE 3-11: **Snip the points off the skewers before using them.**

3. The marble run consists of two towers that hold up rows of slanted tracks. The tracks have a narrow end and a wide end. At the narrow end, the two tracks are just far enough apart to allow the marble to roll along them. At the wide end, the marble is able to slip through to the next level down. To help you figure out the measurements of your system, create a sample set of tracks. Take two skewers. Lay a marble on the work surface. Line up the skewers on either side of the marble. This is the widest point between the tracks. You should be able to lift the skewers up easily, without rubbing the marble. Take one of the short pieces of skewer you cut and glue it across the two tracks to hold them in place, about ¼ inch (6 mm) from the ends of the tracks. The other end of the skewers should be almost touching. You can use the thin

FIGURE 3-12: **Cut some short pieces for crossbars.**

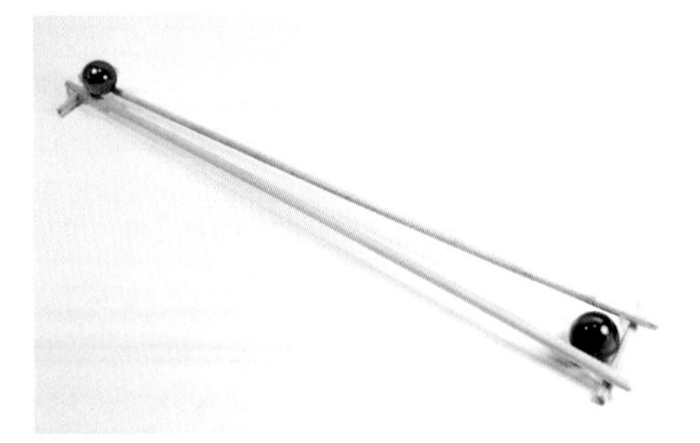

FIGURE 3-13: **Make a sample set of tracks as a guide. The skewers should be angled so a marble can fit between them easily at one end.**

edge of a tongue depressor as a spacer while you glue another short piece across the other end.

4. To make the tower supports, take a skewer and poke the pointed end into a small bead (preferably a bead with a flat bottom). Secure it with a little hot glue. (Use a piece of scrap wood to wipe off any excess glue that leaks out.) Let it cool. Repeat with three more skewers, for a total of four.

FIGURE 3-14: A crossbar holds the tracks in place.

5. Now it's time to assemble the side towers. They are made up of two of the sticks with the beads, with a tongue depressor as a base. To figure out the proper width of the tower, take two of the tower skewers with the beads on the end and lay them down side by side. Fit the wide end of your sample track around them. Spread the tower skewers apart until they touch the inside of the track skewers. Glue a crossbar (short piece of skewer) across the two tower skewers to hold them in place, right above the beads. Check to see that the narrow end of the sample track fits inside the tower. If it does, then glue the bottoms of the beads onto a tongue depressor. If not, adjust everything by carefully peeling apart the glued sticks and

FIGURE 3-15: Squirt a little hot glue into the bead first to hold the skewer in place.

FIGURE 3-16: Glue the sticks with the beads to tongue depressors for support. Attach crossbars higher on one side than the other so the bottom set of tracks tilts down.

regluing them in the proper positions. Repeat with the other tower—but the crossbar for the second tower should be about 1 inch (2.5 cm) higher than on the first tower. Measure the position of the higher end of the track and mark it on the tower in pencil before gluing.

6. To make the bottom set of tracks that will help support the frame, set two skewers inside the tower sticks, resting on the crossbars. Leave about ½ inch (12 mm) of the skewers hanging off either end. Attach the skewers to the crossbars with hot glue. Before going on, roll a marble down the tracks to make sure they're tilted enough—if not, carefully remove the higher crossbar, raise or lower that end of the tracks, then reglue. Next, glue crossbars onto the tops of the tower sticks, making sure that the width of the towers stays the same at the top and the bottom. Connect the tops of the towers by gluing two skewers on the outside of the towers, just below the top crossbars. Test again to make sure a marble can fit between the skewers at the top of the frame.

FIGURE 3-17: The frame of the marble maze (with extra crossbars for later use)

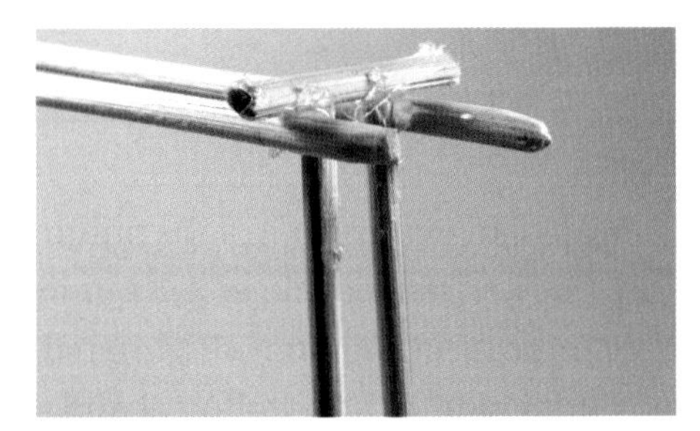

FIGURE 3-18: A close-up of the crossbar on top of one of the towers, and the skewers that connect one tower to the other

7. Now you can add tracks inside the frame to make the different levels. You can control the speed of the marbles by altering the slope of the tracks. The steeper the angle of the tracks, the faster the marbles will roll. The slant will also determine how many sets of tracks you can fit into the framework of the marble run. The sample system in the photos has six sets of tracks. Here are some tips to keep in mind:

- Like the sample track that you made, every set of tracks has a narrower end, where the skewers are glued to the inside of the towers,

HOW TO REMOVE HOT GLUE

To fix a section of your marble run after it has been hot glued, let the glue cool. Then carefully peel away the dried glue that holds the pieces together. Remove as much of the old glue as you can without damaging other parts of the marble run. Move the pieces to their new positions and test them to make sure everything works. Then use more hot glue to reattach the parts. If there is still old glue in the way, you can gently soften it by touching it with the side of the hot metal nozzle on the glue gun. Hold the pieces together until the glue cools and sets. And be very careful not to get hot glue on your skin!

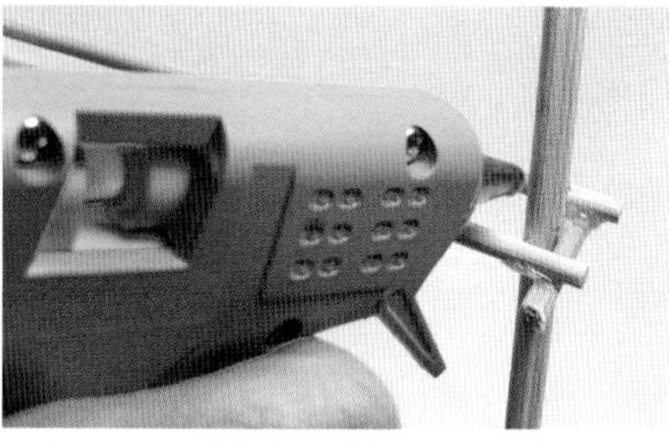

FIGURE 3-19: **Hot glue is quick and easy to use, and not that difficult to remove.**

and a wider end where the tracks are glued to the outside of the towers.

- The narrower end of the tracks is always higher than the wider end.
- The tracks will zigzag—on one level the tracks tilt down to the left, the next set tilts down to the right.
- The wide ends and the narrow ends of the tracks alternate—the wide end of one level sits above the narrow end of the level below it. The marble will drop from the wide end of one level to the narrow end of the level below it.
- Use your sample set of tracks to figure out where the high and low ends of your track will be connected to the

FIGURE 3-20: **A second level has been added above the bottommost tracks. (On this version, slanted braces were added to guide the marbles. They were later removed and replaced with guard railings.)**

towers. You can vary the slope of each level, or just make them all the same. If you want your marble to bang into or bounce over noisemakers, tilt the track so it is a little steeper. That will give your marble more speed and keep it from getting stuck.

- You can also add speed by increasing the distance the marble has to drop between one level and the next. The drop should be at least ½ inch (12 mm), but you can make it as much as 2 inches (5 cm).
- Be sure to leave enough room between levels so you can hang noisemakers like keys and washers.

MAKING MUSIC OUT OF NOISE

Part of the challenge of building a Musical Marble Run is getting everything to work right. The other part is getting it to create clacks, bangs, and bongs that sound musical! Remember that you can "program" your marble run as you design it. First, you can control the tempo, or speed, of the music by controlling how fast marbles roll along. You can also control when sounds occur and in what order by the way you arrange the various noisemakers. As you experiment with different placements, keep in mind what kind of notes and rhythm you want your marble run to play. That's how you turn a toy into a musical invention!

8. You already have the bottommost tracks in place. To build the next level up, take your sample set of tracks and insert it between the towers. Follow the guidelines in Step 7 to position it correctly, then mark the position of the ends on each tower. Remove the sample and glue crossbars onto the towers at the marks. When they are dry, rest two skewers on the crossbars. At the *lower* crossbar, glue the

FIGURE 3-21: More levels being tested

skewers to the *outside* of the tower. At the *higher* crossbar, glue the skewers on the *inside* of the tower.

9. Test the new level with a marble. Then continue adding on levels the same way until you reach the top of the frame.

FIGURE 3-22: A close-up of the wide end of a set of tracks, with guard railings above it

10. To keep the marbles from falling off the sides of the tracks, make "guard railings" by gluing skewers on the outside of the towers, a little above each track. (They should be about half as high above the tracks as the marbles are tall.)

11. Once your marble run is operating smoothly, you can start to add noisemakers. These can be attached to crossbars glued to the underside of the tracks where needed. Be sure to test them before you attach them permanently to make sure they don't slow or stop the marble along the track. Here are some ideas:

    - Hang a bell over the track. If needed, make the clapper longer so the marble hits it as it rolls underneath.
    - Hang keys, metal washers, or recycled mini wind chimes from above.
    - Make a clackity "raft" with mini craft sticks. Line up the sticks, attach two sticks across the raft on both sides, above

FIGURE 3-23: A bell is hung from an overhead crossbar. A metal washer attached to the clapper makes it easier for the marble to set it ringing.

FIGURE 3-24: Mini wind chimes are hung close together so one hits the next, making them all ring.

and below, and tie these sticks together with string. Hold the whole raft in place by tying another craft stick to a crossbar loosely fastened across the tracks.

- Attach a string of beads between the tracks just high enough for the marble to bump along as it rolls downhill.

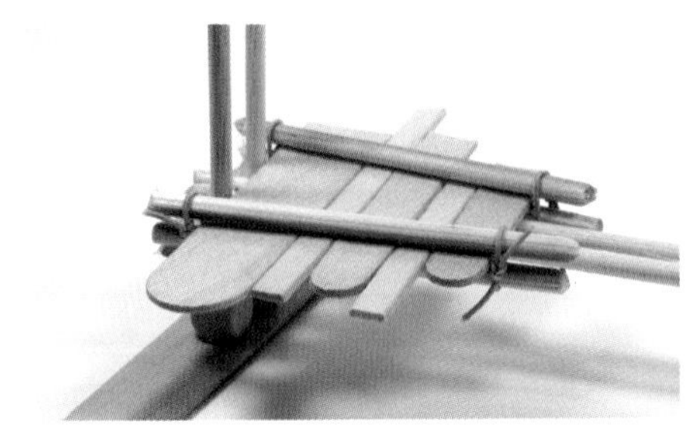

FIGURE 3-25: **For maximum sound, these sticks are held together with sticks and string, like a raft, and tied to a crossbar underneath the tracks.**

- Save large bumpy noisemakers—such as large wooden beads or metal bottle caps—for the spot right below the drop from one level to another. That will give marbles enough extra momentum to get over the rough spots.

FIGURE 3-26: **Marbles make a raspy sound as they roll over a string of bumpy wooden beads.**

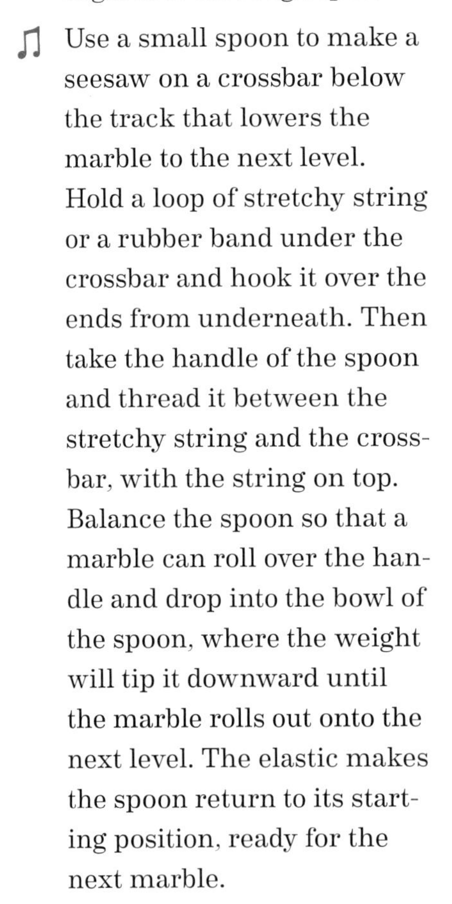

- Use a small spoon to make a seesaw on a crossbar below the track that lowers the marble to the next level. Hold a loop of stretchy string or a rubber band under the crossbar and hook it over the ends from underneath. Then take the handle of the spoon and thread it between the stretchy string and the crossbar, with the string on top. Balance the spoon so that a marble can roll over the handle and drop into the bowl of the spoon, where the weight will tip it downward until the marble rolls out onto the next level. The elastic makes the spoon return to its starting position, ready for the next marble.

FIGURE 3-27: **Large wooden beads are tied together and then tied to the tracks. Marbles drop onto them and roll down onto the tracks.**

♫ Be sure to end the marble run with some kind of noisemaker, such as a wind chime laid crossways across the railings.

FIGURE 3-28: This spoon works like a seesaw that gently lowers marbles to the level below.

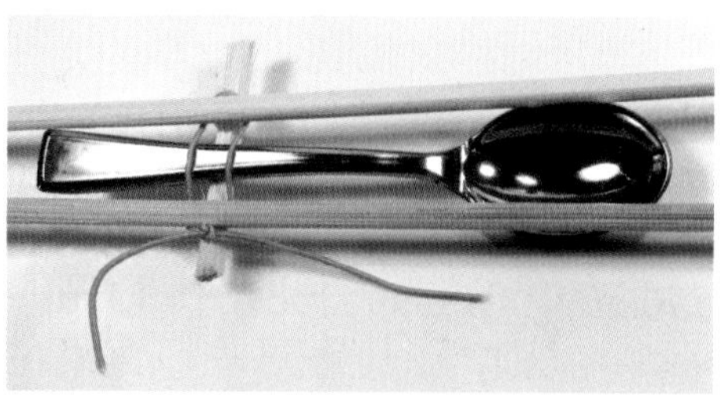

FIGURE 3-29: Elastic string is used to tie the spoon to a crossbar at its pivot point.

FIGURE 3-30: A close-up of a marble about to roll off the spoon. The bottle cap is wedged into place, not glued.

FIGURE 3-31: A metal chime is tied loosely to a crossbar at the very end of the marble run.

GETTING THE BEST SOUND FROM YOUR MARBLE MAZE

When adding noisemakers of various kinds to your marble run, remember what you learned about standing waves on an idiophone: in order to get the best sound, you should fasten the vibrating piece where the nodes—the points that stay still—occur. For wooden and metal bars, the nodes for the fundamental frequency (the main pitch you hear) are a little less than one quarter of the way in from either end. If it's not easy to fasten the bar at that point, attach it as loosely as possible so it can vibrate freely.

Music Box Engineering

Music boxes are known for their tinkly sound. But wind-up music boxes are also finely tuned machines. Turn the key to tighten a spring, open the lid to release the catch that lets the spring unwind, and gears and flywheels start to move the mechanism at just the right speed. The music is produced by a rotating cylinder studded with tiny pins. As the cylinder (known as the *drum*) slowly spins, the pins pluck the teeth of a metal comb. The teeth on each comb are different lengths and thicknesses, depending on the notes needed for the melody each box plays. Every part of a music box takes artistry and expertise to create.

FIGURE 3-32: The inner workings of an antique music box. You play it by winding up the spring (in the round casing at the upper right). This turns gears that make the drum with the small pins rotate, plucking the tines of the metal comb as it slowly spins.

Music boxes also have inspired musical inventors to get creative. Like thumb pianos, the teeth can be made out of any material that makes an interesting sound when it is plucked or tapped. Our version uses assorted wire springs, but you can try other kinds of noisemakers as well.

MUSICAL INVENTORS: KOKA NIKOLADZE

This Experimental Music Box was inspired by Koka Nikoladze of Norway (*nikoladze.eu*), who builds hand-cranked Beat Machines—music boxes that use objects like forks and plastic rulers instead of a metal music box comb. He also makes computerized versions using programmable Arduino microcontrollers. His other musical inventions include a MacBook Pro that you play with a bow. He has said that his goal is to make a bassoon you can read your email on.

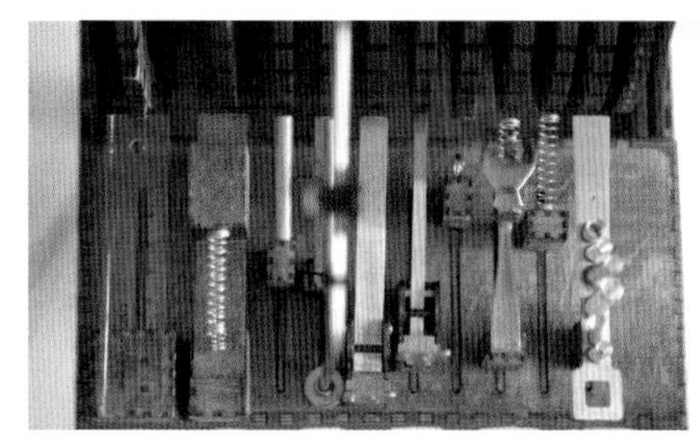

FIGURE 3-33: Koka Nikoladze uses springs, rulers, forks, and other objects as the comb in his music boxes. The wooden disks have movable pins that let him program the tune.

Project: Experimental Music Box

Materials

Wooden box, such as a small cigar box

Wooden board that fits snugly inside the box when it stands on end; it should be about 1 inch (2.5 cm) in depth and at least 1 inch (2.5 cm) higher than the sides of the box.

5 or more wooden coffee stirrers (depending on the width of your box)

5 or more assorted wire springs (available in hardware stores, or reuse old springs from ballpoint pens and other household items)

4 pencils or other stiff narrow rods, such as dowels or chopsticks

Rubber bands

6 wooden clothespins (or other kinds of clips)

Wooden cylinder at least 6 inches (15 cm) longer than the box is wide, such as a rolling pin (you can sometimes find them in dollar stores) or a piece of a closet rod

Assorted wood screws

Pushpins

(Optional) craft stick or tongue depressor

(Optional) peel-and-stick craft foam

(Optional) tape

Tools

Small pliers for bending the wires in the springs

Screwdrivers (that match the heads of the screws you are using)

Saw (if needed to cut board and rods)

Safety Warning: Children should get adult help cutting materials like wooden boards and rods.

Note: Because it's experimental, this project uses quick-and-dirty construction techniques that are easy to modify but not very permanent. Can you think of ways to improve the design to make it play better and last longer? Test your ideas to see if they work!

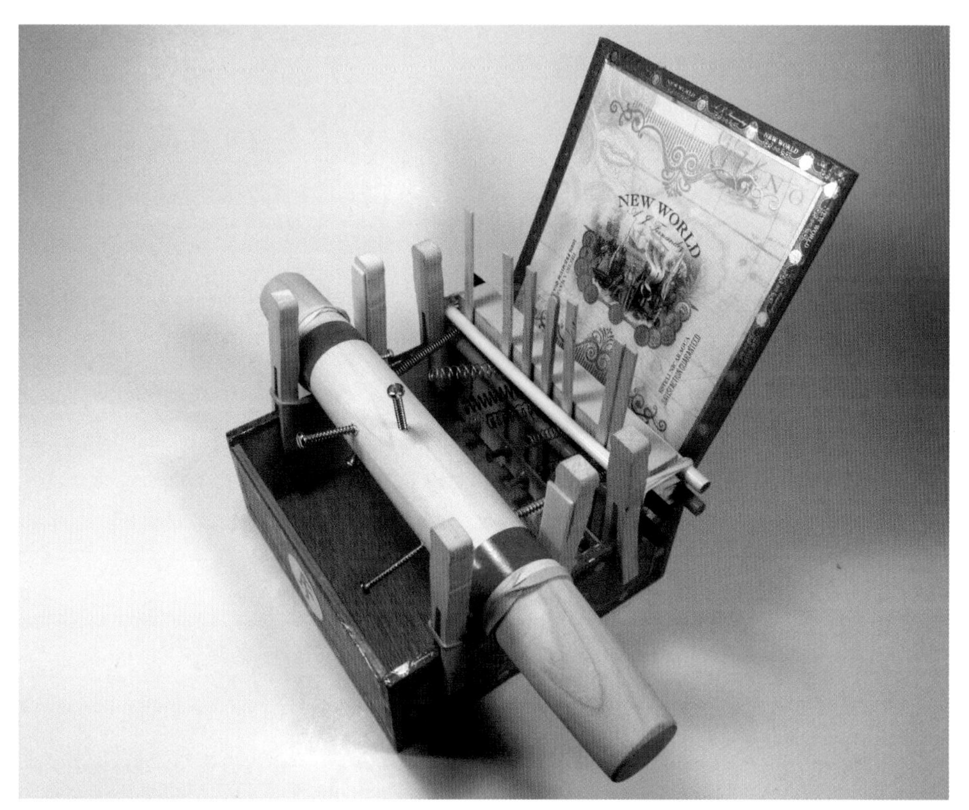

FIGURE 3-34: **This prototype Experimental Music Box is easy to change around.**

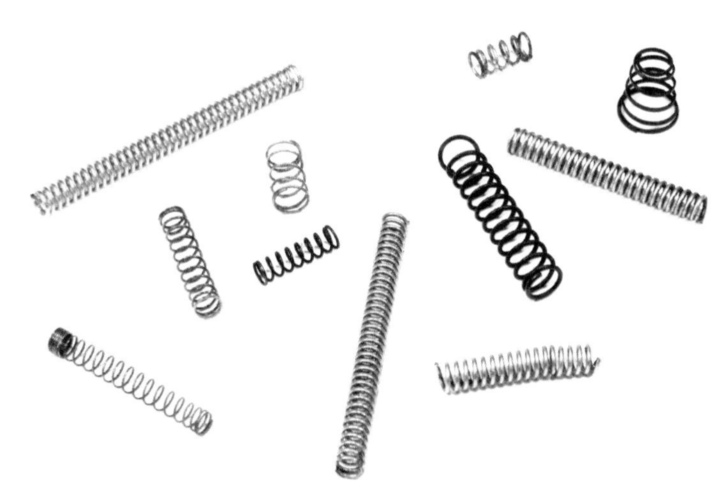

FIGURE 3-35: **Assorted springs, some taken from household items like ballpoint click pens.**

FIGURE 3-36: The same kind of cigar box used to make a guitar in Chapter 1 can be used for a DIY music box.

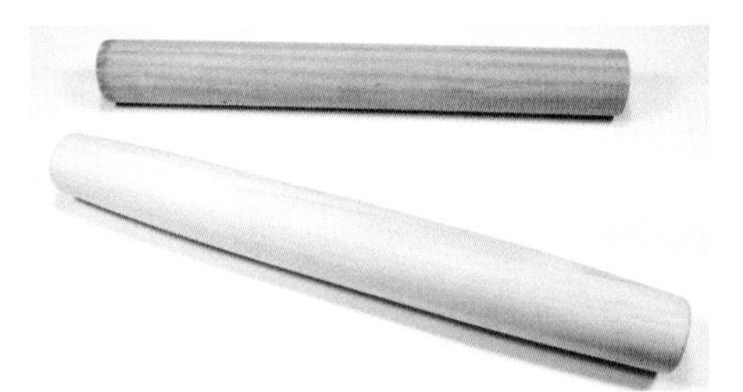

FIGURE 3-37: Two options for a music box drum: a piece of closet rod (top) or a rolling pin (this one from a dollar store).

1. The "comb" of the Experimental Music Box follows the same basic design as the Thumb Piano earlier in the chapter. You will attach coffee stirrers to a wooden board using rods and rubber bands. However, in this case, the coffee stirrers are simply used as anchors for the springs. The board then fits into the box (which also holds the drum you turn to pluck the springs). The first step is to figure out where to place the springs. To get started, take the board and stand it on end inside the box. Then lay a pencil (or other rod) so it rests on the sides of the box next to the board. Draw a line above the pencil across the front of the board.

FIGURE 3-38: Mark the height of the pencil that serves as a rod to hold the tines on.

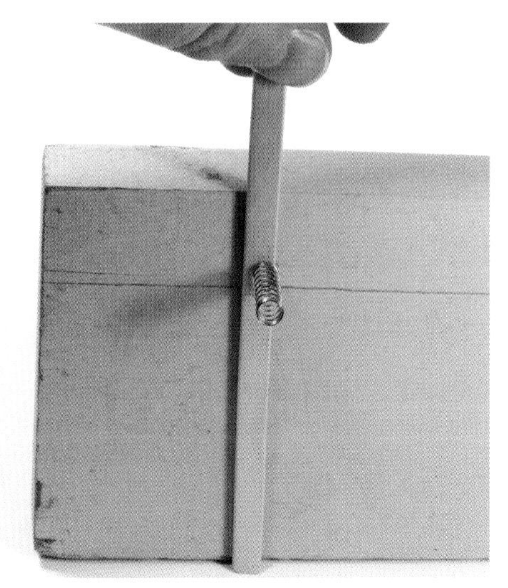

FIGURE 3-39: Mark the height of the line on the board on the coffee stirrers. The springs will be attached above the line as shown here, so there is room for the rod below.

2. Take the board out of the box, still standing on end. Take a coffee stirrer and stand it up against the front of the board. Mark the height of the line on the coffee stirrer. Now attach a spring to the coffee stirrer, just above the line. To do this, squeeze the stick between the coils of the spring as close to the bottom as possible. Turn the spring (like unscrewing a bolt) until just the bottom tip of the wire is below the stick. Take the pliers and pull the tip until it wraps around the far edge of the stick. This will hold it in place. Repeat with the other springs.

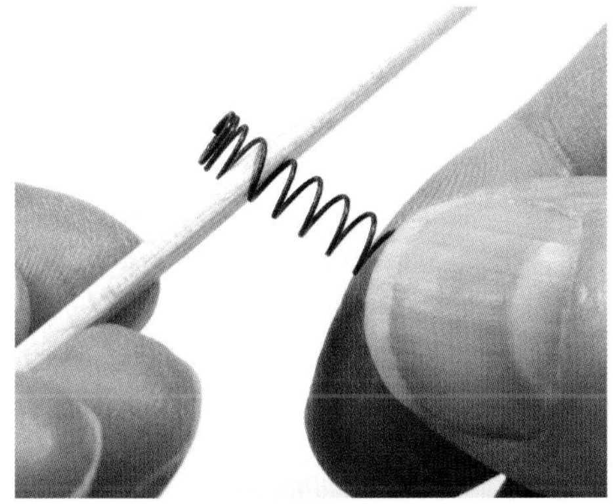

FIGURE 3-40: If the coils at the end of the spring are too tight to slide the stick between them, start at a point where they are wider and turn the spring (like loosening a screw) to move it down.

FIGURE 3-41: The end of the spring underneath the coffee stirrer

FIGURE 3-42: Use a pair of pliers to stretch the end of the spring up and around the side of the coffee stirrer to help hold it securely in place.

3. Lay the board down and line up the coffee stirrer sticks with the springs attached so the springs are sticking up. They should be evenly spaced. Leave room on either end. If any of the springs are angled toward the spring next to them, adjust the spacing as needed. Place a pencil or other rod on top of the sticks, next to the springs. Place another pencil on the back side of the board, and attach them with rubber bands, just like you did with the Thumb Piano. Do the same with the other pair of pencils on the other side of the springs. When you're done, trying plucking or scraping the springs to see if they are tight

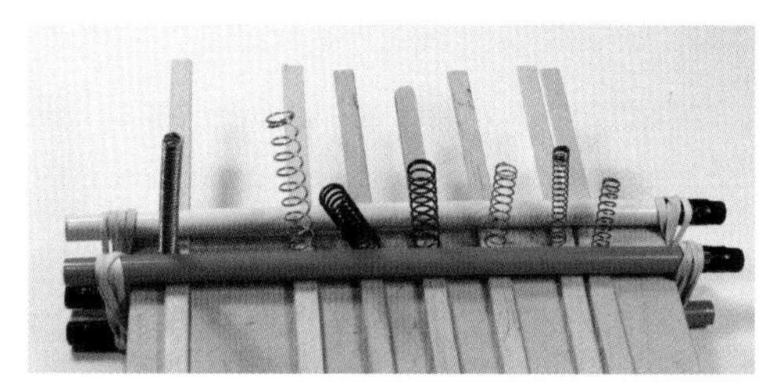

FIGURE 3-43: The coffee stirrers with the springs are attached the same way as the tines of the Thumb Piano.

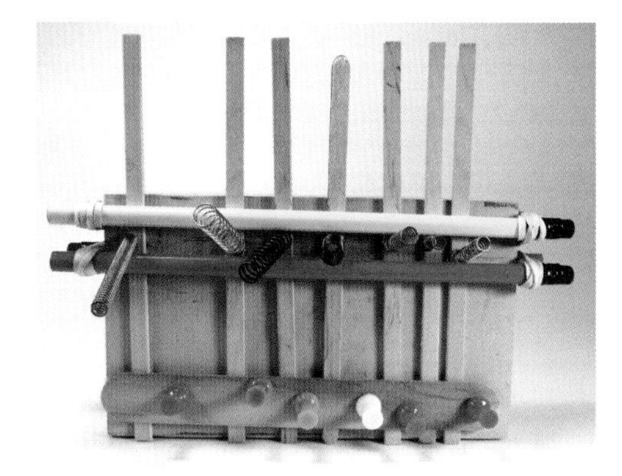

FIGURE 3-44: If needed, add a tongue depressor and some thumbtacks to help hold the coffee stirrers in place.

enough to make a good tone. If the pencils curve a bit in the middle and make the middle stick loose, wedge a shim (such as a piece of paper or a second coffee stirrer) behind it. If you want to secure the ends of the sticks, press them down with a tongue depressor held in place with pushpins.

4. Put the board back into the box. Push it all the way back against the side of the box so the springs stick out toward the middle. Take two of the clothespins and clip them over the sides of the box to hold the board in place. To make a tracing of the springs to help with positioning the pins on the rotating drum, hold a piece of paper flat under the springs and trace around them.

5. Now it's time to make the drum. You will use screws as pins to pluck, scrape, or tap the springs to get them to make a sound. To begin, take your wooden rod or rolling pin and lay it across the top of the box, leaving a little space between the rod and the longest spring. Use more clothespins as guides to hold the rod in place. Make sure at least one end of the drum is sticking out far enough past the clothespins for you to hold onto it in order to rotate the drum. Wrap some tape around the rod.

6. To keep the rod from sliding side to side, wrap rubber bands around the ends of the rod just outside the clothespins to make a little ridge as guides. Wrap some tape around the rod to keep it from rubbing against

FIGURE 3-45: The block with the springs attached is inserted into the box. Clothespins hold it in place.

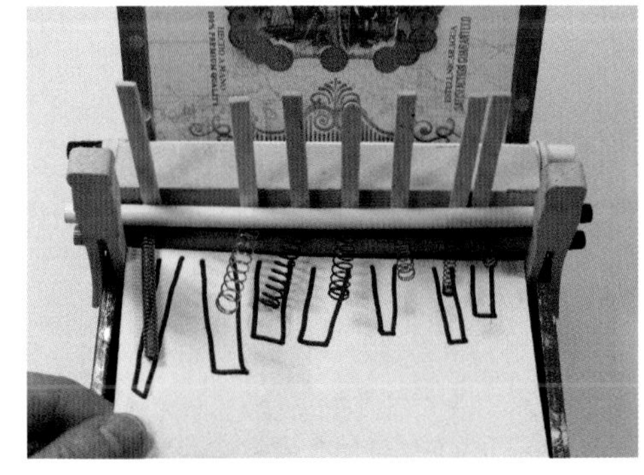

FIGURE 3-46: Make a tracing of the springs to help find screws that are the right length to reach from the drum.

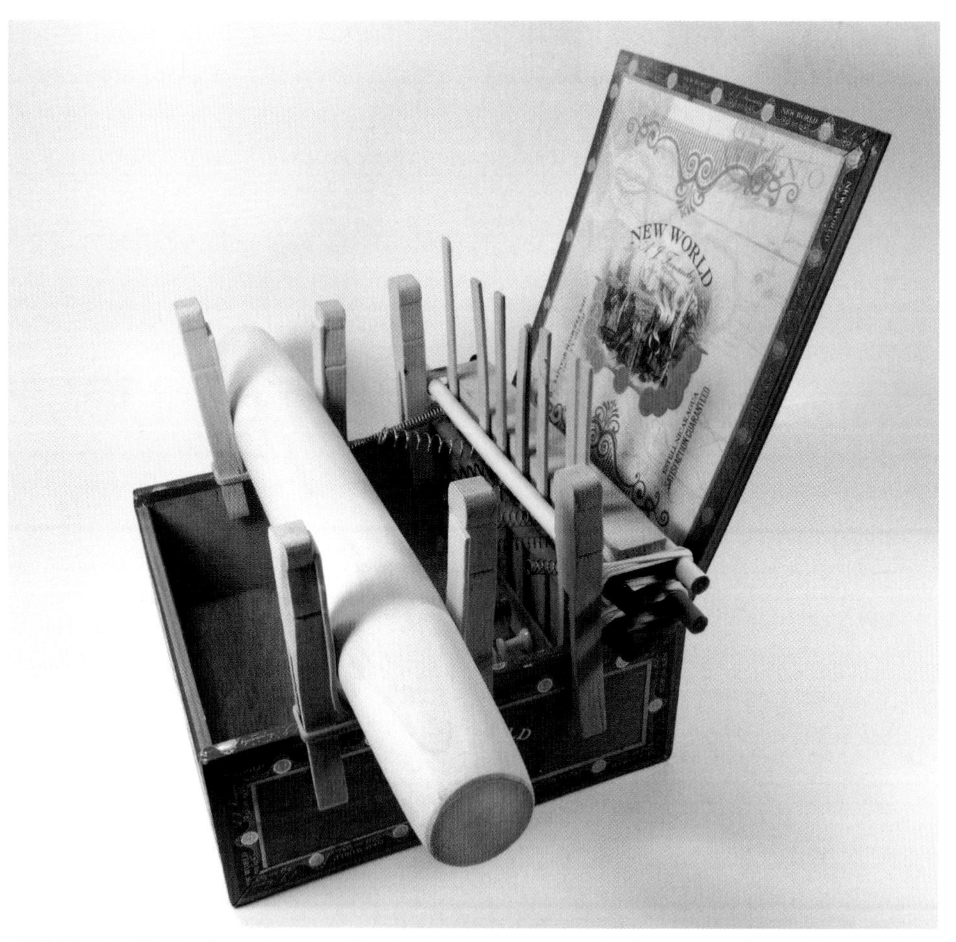

FIGURE 3-47: The drum is placed in the box. Leave room for the screws that will pluck the springs as the drum is turned.

the clothespins or the side of the box. (See Figure 3-50, later in this activity.) To prevent the rod from rubbing against the box, loop loose rubber bands around each pair of clothespins to hold them together. Slide the rubber bands down until they are just above the side of the box. Let the rod hang in them like slings. You can also stick a strip of peel-and-stick foam tape along the inside of the clothespins, in a "U" shape.

7. Now it's time to attach the pins. For the first pin, find a screw that is a little bit longer than the distance between the rod and the first spring. *Important: The head of the screw should be wider than the spring.* Make a mark where the screw should go—hold a stick straight across from the spring to the rod to help you line them up. Take the drum out and use a pushpin to make a starter hole at that spot. Then take the screw and a screwdriver and carefully turn the screw a few times into the wooden rod. Test the screw by replacing the drum and turning it. The screw should touch the spring to make a sound, but not get caught in the coils. If you need to adjust the distance, use the screwdriver to turn the screw further into or out of the drum.

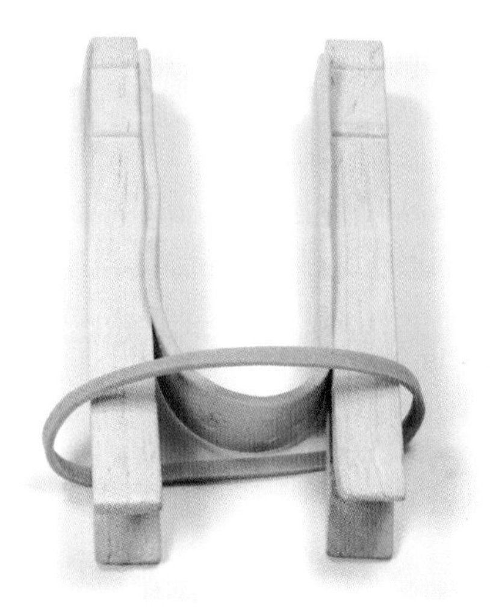

FIGURE 3-48: The drum holder shown separately. Two clothespins hold the drum in place on either side. A strip of peel-and-stick craft foam cushions the drum to prevent it from banging against the clothespins or the box. A rubber band helps hold the pins to keep them from slipping.

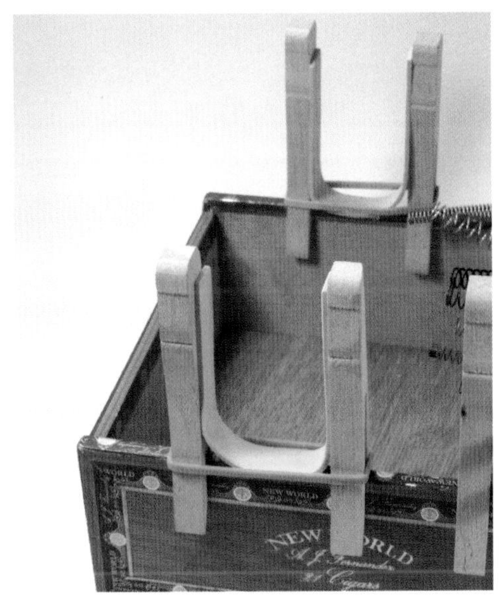

FIGURE 3-49: The drum holders fastened to the box

8. Mark the spots for the remaining pins. If you want to play two or more notes at the same time, line up the pins for those springs in a row straight across the drum. If you want to play the notes separately, place them at different spots around the rod. Insert the screws as you did in Step 7, testing after you add each one before you go on to the next screw.

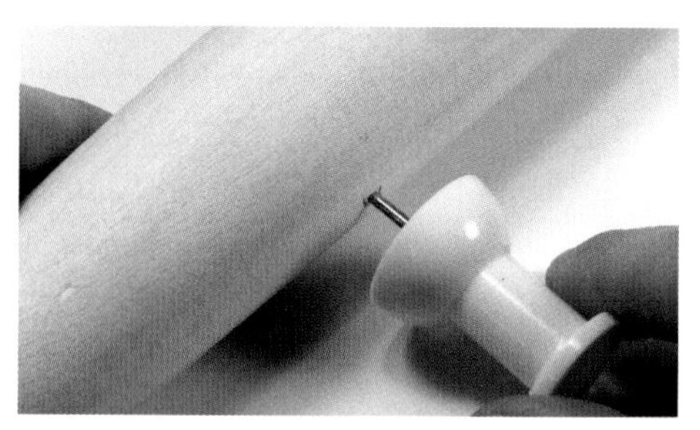

FIGURE 3-50: Use a pushpin (this one is oversized) to make starter holes for the screws.

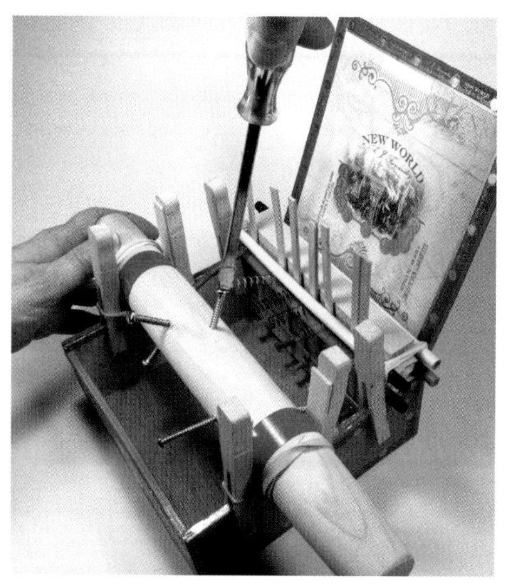

FIGURE 3-51: Adjust the screws so that they bend the springs down without getting caught in the coils.

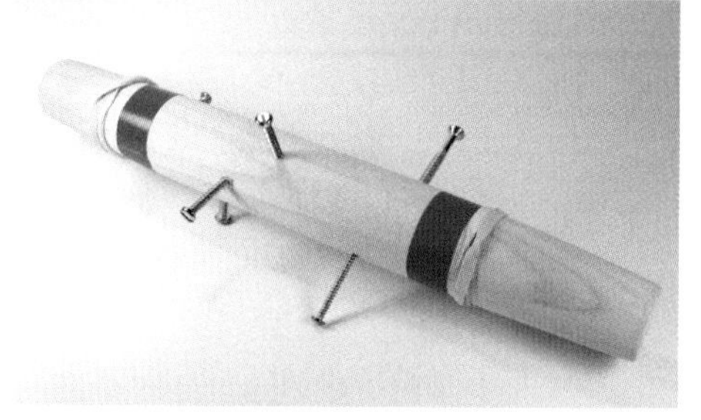

FIGURE 3-52: The finished drum shown separately. The electrical tape reduces rubbing, and the rubber bands fit on the outside of the drum holders to keep the drum from sliding side to side.

FIGURE 3-53: Overhead shot of the finished music box. You play it by turning the ends with your hands.

Project:
Origami Paper Popper

> **Materials**
>
> **1 sheet of thin paper (like lined loose-leaf or yellow legal pad paper)**

Anything that can create pressure waves in the air can produce a sound. And anything that produces a sound can be used to make music—even sheets of paper. An Origami Paper Popper makes an explosive *crack!* when you flick it. It is a type of free aerophone, like the bullroarer you swing around your head from Chapter 1, which makes the air around it vibrate directly. Although typically used for impressing friends on the school playground, you can also use your Origami Paper Popper as a percussion instrument in your Musical Inventions band.

1. Lay the sheet of paper in front of you horizontally, so it is longer from side to side. Take the top edge and fold it down so it meets the bottom

edge. Press the crease flat, so it is nice and sharp and divides the paper evenly in half. Then open the paper again.

2. Take the top edge and neatly fold it down to the crease you just made across the middle of the paper. Make these folds as precise as you can so the folds are as thin and as exact as possible. Sharpen that crease and repeat—that is, fold the new top edge down to the middle crease.

FIGURE 3-54: Fold the top edge down, then open the paper up again.

3. The top half of the paper is now folded up several times. Take the folded top edge one more, and fold it down along the original middle crease. Then fold the folded top edge one last time. You now have a big fat folded strip across the top of the remaining paper.

FIGURE 3-55: Fold the top edge down so it meets the fold across the middle and leave it folded.

4. Flip the paper over so the folded-over flap is against the table. Take the right edge and fold it over so it meets the left edge. Make the crease as smooth and sharp as possible.

FIGURE 3-56: Fold the folded top edge down to the middle crease once more.

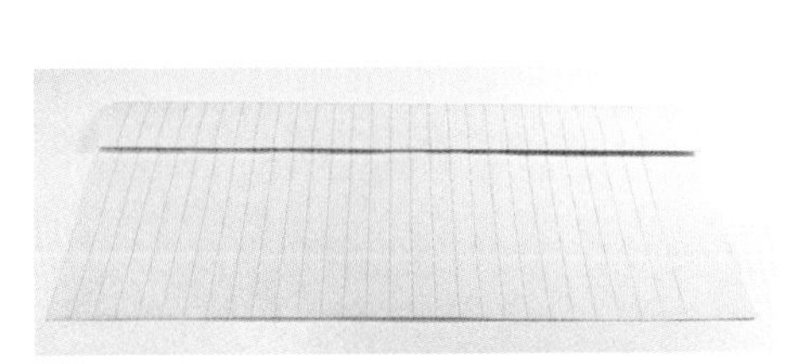

FIGURE 3-57: Fold the folded top down along the original crease line across the middle of the paper.

FIGURE 3-58: Make one last fold. You will have a big fat folded flap across the top of the paper.

5. Finally, "load" the popper so it is ready to go. To do so, turn the paper so the flap is on the left. (*Note:* If you are left-handed, reverse the left and right directions in this step.) Hold the paper by the top corners. Carefully pull the upper right-hand corner out and down, so that the crease opens up into an upside-down V. The sides of the V should be fat and rounded. Stop when just the tip of the right-hand corner is still visible. Move your right hand to hold onto the bottom corners of the V. You should now be holding the popper like a backward paper airplane, with the bigger end of the open rounded "tunnels" facing forward, and the point facing the back.
6. Quickly flick the popper downward. The rounded parts should pop open with a loud *snap*! If it doesn't work the first time, try opening the tunnels up a little with your fingers to give them a head start.
7. To reload the popper for another pop, push the now-opened flap back into the position it was in at the start of Step 6. Your Origami Paper Popper should be good for several snaps!

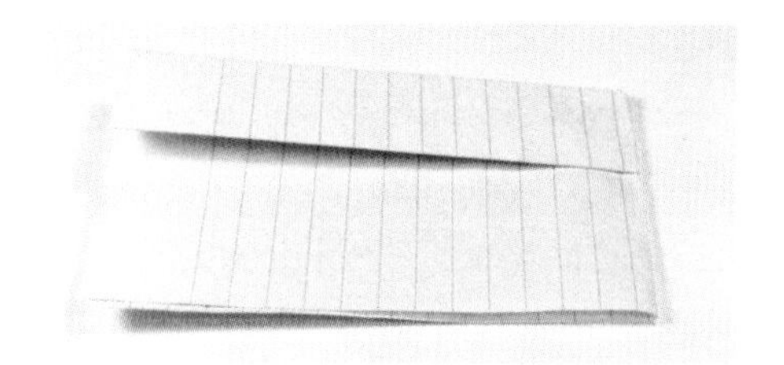

FIGURE 3-59: With the flap side down against the table, take the right edge and fold it over so it meets the left edge.

FIGURE 5-60: Turn the paper so the flap is on the left. Hold it by the top corners.

FIGURE 5-61: Start to pull the right hand corner out and down. This will make the flap start to open up.

FIGURE 5-62: Keep pulling until the flap opens up into two cone-shaped tunnels. Then move your right hand and grab the two points that are now sticking out at the bottom right.

FIGURE 3-63: A view from the front of the Origami Paper Popper shows the two tunnels formed by the opened flap.

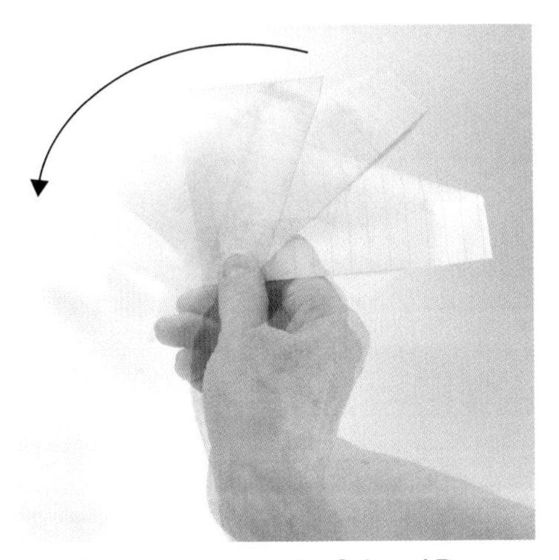

FIGURE 3-64: To play the Origami Paper Popper, start by holding it like a backward paper airplane. Bend your wrist back and fling it quickly forward. The rounded openings will snap open with a loud crack.

FIGURE 3-65: To reset the popper, push the center of the V back in with your finger.

Project: Flapping Paper Strips

Materials

- Thin paper (such as loose-leaf or tracing paper)
- Scissors
- Tape
- Small table fan
- (Optional) homemade membranophone drum (see Chapter 2) or tambourine

Paper isn't just for popping. You can use paper to create quieter, more subtle sounds by crinkling it or letting it flap in the breeze.

Safety Warning: Be careful not to get your fingers near the fan's blades when it's spinning. Turn the fan off before you add or adjust the paper.

1. Cut one or two thin strips of paper. Tape the paper to the front of the fan.
2. Turn on the fan to see if the paper flaps and makes interesting sounds. Try different widths and lengths, or try making cuts in the end of the paper to create a fringe.

Extension:

- If you have a drum or tambourine, tape a strip of paper to the drumhead and aim the fan at it. See if it will create a varying beat as it flaps against the drum.

FIGURE 3-66: This self-playing "instrument" takes only a fan and some strips of paper.

MUSICAL INVENTOR: PIERRE BASTIEN

French composer Pierre Bastien (*www.pierrebastien.com*) uses fans to turn strips of paper into musical instruments. He programs groups of fans to turn on and off and make paper strips tap against the heads of paper drums and tambourines, creating hypnotizing rhythms. Bastien's other musical inventions include *Mecanologie*, which uses motorized construction toys to build simple musical robots using objects from daily life. His creations make teapot lids clatter, strum the teeth of a comb, scrape toothbrushes back and forth across a set of false teeth, swing a letter-scale back and forth, open and close a creaky pair of scissors, and bang a hammer in a syncopated rhythm against a block of wood.

FIGURE 3-67: Pierre Bastien's piece *Paper Harp* is made up of a small fan and a paper cut into strips that gently pluck or hit the strings of an African *kundi* harp.

4

Eerie Electronic Music

Hack old devices to create weird sounds, and invent unusual controls for your own computer compositions.

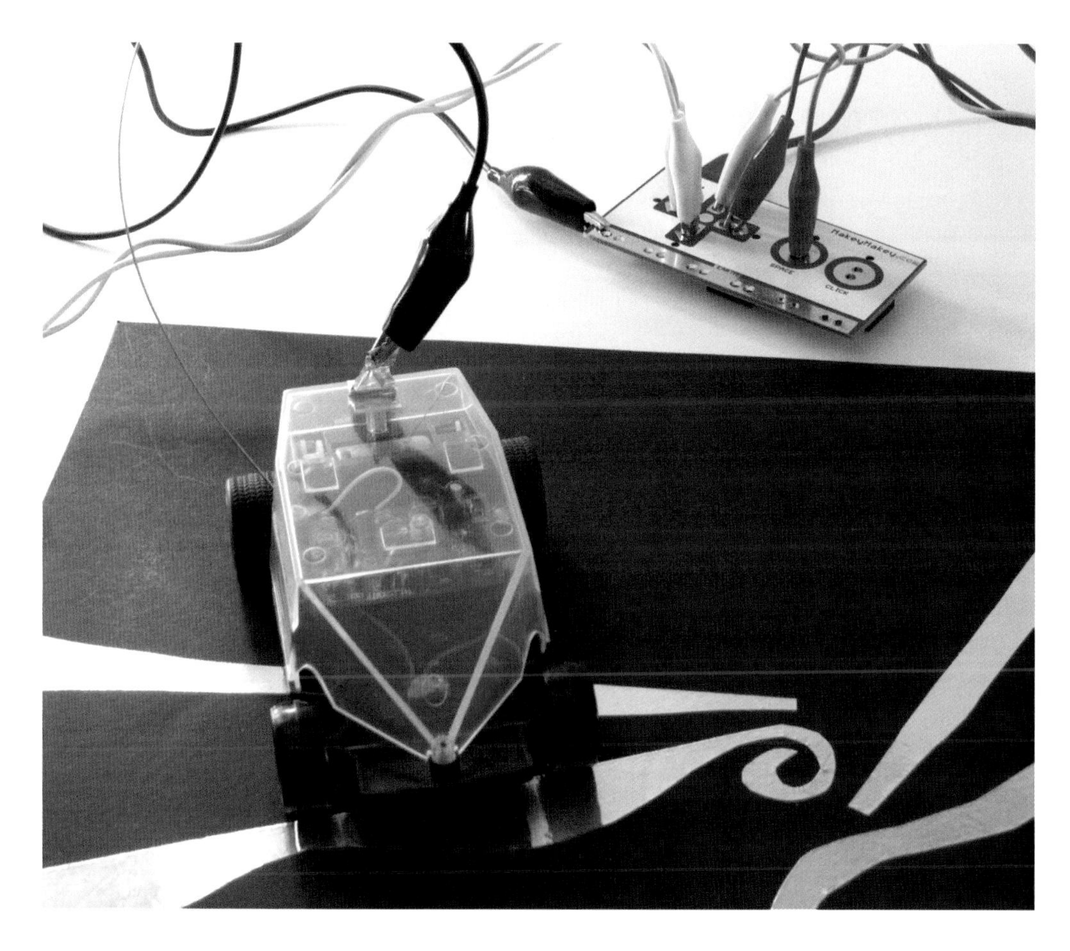

So far in this book, you've learned how to change the sound produced by an instrument—at least for acoustic instruments. *Acoustic instruments* produce sound waves when they vibrate. Change the size, shape, or material the instrument is made out of, and the instrument will vibrate differently. Build an instrument that's big and heavy, and it will vibrate slower and produce lower notes. Build it small and light, and it will vibrate faster and play higher notes. If you make it out of wood, it will have overtones that add richness to the tone. If you make it out of metal, it will play a sharp, clear note.

But as you're about to discover, some instruments don't follow the rules. *Electronic instruments* (which fall into the category of electrophones) create vibrations using electrical signals. And those electrical signals can be shaped to produce different notes and tones, in the same way the body of an acoustic instrument shapes sounds. An instrument that generates electrical signals and shapes them to produce musical sounds is called a *music synthesizer*. (*Synthesize* means "put together.")

Music synthesizers have been around for over a century. In the late 1890s, an instrument called the telharmonium sent musical tones in the form of electrical signals over special phone wires. In the 1920s, the theremin made music using radio waves. By the 1960s, synthesizers got smaller and easier to use. Today, synthesizers (or synths for short) are used in all kinds of music. They can create drumbeats, crazy harmonies, and otherworldly sounds. And electronic instruments can be programmed like computers to play certain sounds, patterns, or even entire songs. In fact, music software can turn your computer into an instrument!

Some electronic instruments look a little like acoustic instruments. They may have keyboards like a piano, strings like a guitar, or a mouthpiece like a wind instrument. In addition, they may have a control panel filled with switches and knobs that let you modify the sound in endless ways. But electronic instruments can take all kinds of interesting and artistic forms. The only thing that matters is what electrical components are connected to them.

In this chapter, you'll get to play around with electrical circuits and explore ways to produce strange and wonderful sounds with everyday devices.

MUSICAL INVENTOR: ROBERT MOOG

In the 1960s, a scientist named Robert Moog revolutionized the pop music landscape with his music synthesizer. Moog got his start as a teenager, building and selling updated versions of the theremin. (See the DIY Theremin project later in this chapter.) After earning a Ph.D. in engineering physics, Moog started his own company to work on designing new electronic music devices. Moog synthesizers allowed artists to create sounds that had never been heard before. The possibilities became clear in 1968, when a classical album called "Switched on Bach" by Wendy Carlos used a Moog synthesizer to reimagine the keyboard music of Baroque composer J. S. Bach in a new and fun way. The album became a surprise hit, and Moog became a household word. (See "Playing Around with Synthesizers" later in this chapter to try an online simulation of a Moog synthesizer.) After Moog died in 2005, his family created the Bob Moog Foundation (*moogfoundation.org*) to share the history of his work and introduce students to the science of sound through music synthesizers.

Where Electronic Music Gets Its Spark

How does an electronic instrument produce sound? When you play an acoustic instrument and cause vibrations, you are making tiny air molecules shake back and forth. Areas of high pressure (with lots of molecules packed in) and low pressure (with fewer molecules spread out) travel away from the instrument in waves. These pressure waves carry the sound energy from the instrument through the air to your ears.

Synthesizers create similar pressure waves using electricity. *Electricity* is the flow of energy in the form of *electrons*, which are particles found in atoms, the building blocks of molecules. A synthesizer usually contains some kind of generator that causes electrons to vibrate. The term for vibrations in electrons is *oscillation*. Oscillation creates pressure waves that can be used to power speakers or earphones that translate the electrical energy into sound waves you can hear. The measure of the pressure of electrons in a system is its *voltage*.

To shape the electric signals and produce different sounds, synthesizers use several techniques. One is to generate multiple signals and blend them to produce new patterns. Remember how a sound wave interferes with itself when it bounces back from the end of an instrument? Waves of different frequency can be electronically added together in the same way. Another technique used by synthesizers is to filter some of the harmonics that make up a tone. Making different overtones stronger or weaker can change the timbre of a sound.

Just like with sound waves, you can use drawings called *waveforms* to represent how electrons oscillate. The line of the waveform goes up and down to show you changes in voltage over time. The shape of the waveform can tell you what the timbre of the sound is like. Here are some types of waveforms you can generate on a music synthesizer:

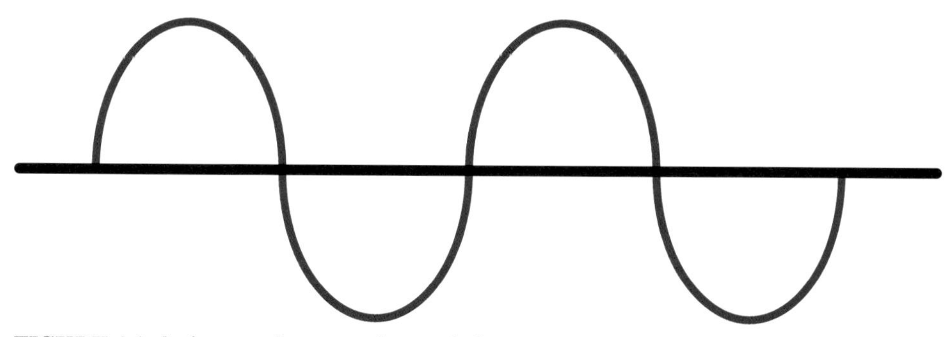

FIGURE 4-1: A *sine wave* has smooth rounded curves. It represents a pure tone with no harmonics, like a whistle.

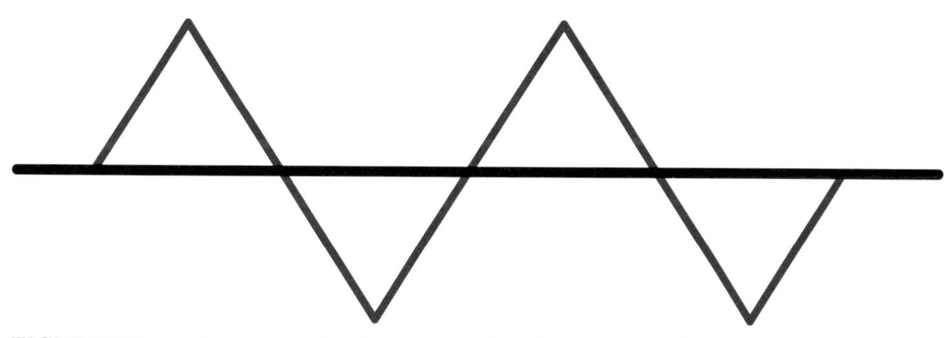

FIGURE 4-2: A *triangle waveform* is less smooth and pure than a sine wave. It sounds something like a flute.

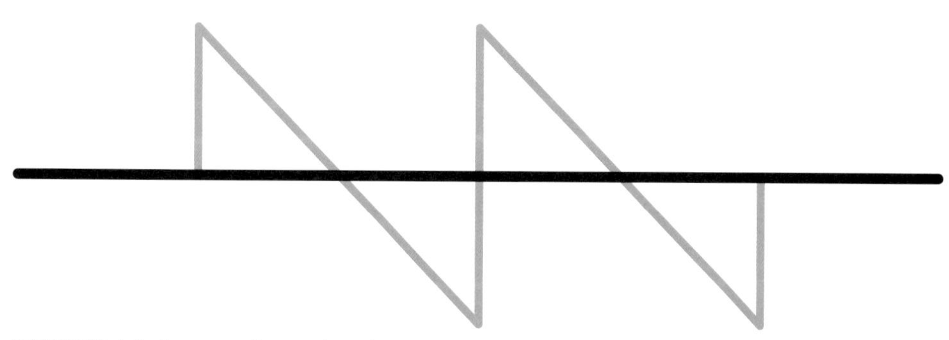

FIGURE 4-3: A *sawtooth waveform* looks like a triangle that's leaning over, with a long build-up and a quick drop. It is bright-sounding, like a string or brass instrument.

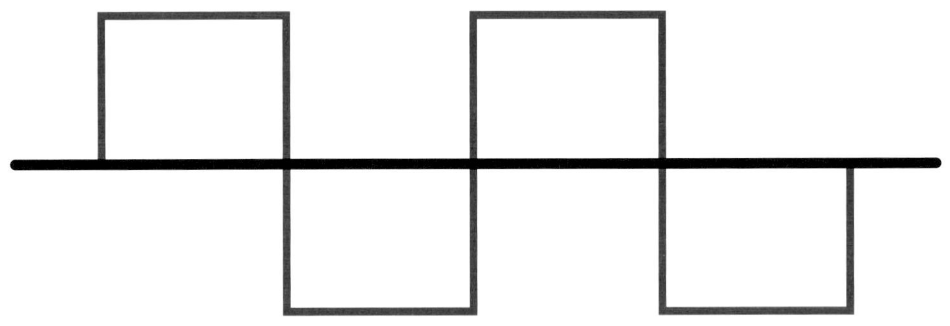

FIGURE 4-4: A *square waveform* has noticeable harmonics. It sounds hollow and woody, like a clarinet.

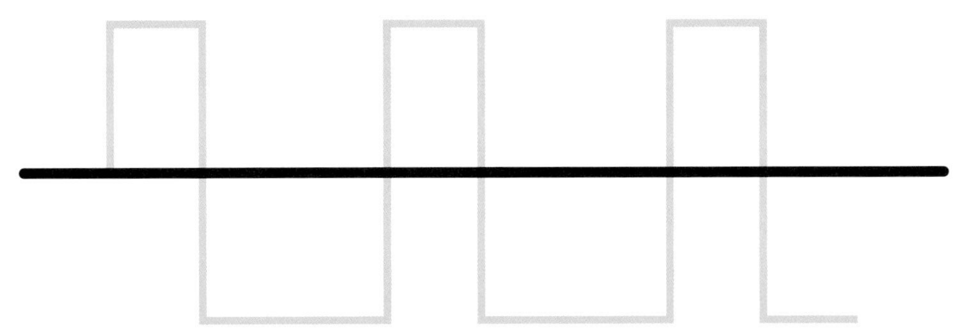

FIGURE 4-5: A *pulse waveform* looks like narrow squares separated by large gaps. It's more nasal than a square waveform and sounds like an oboe.

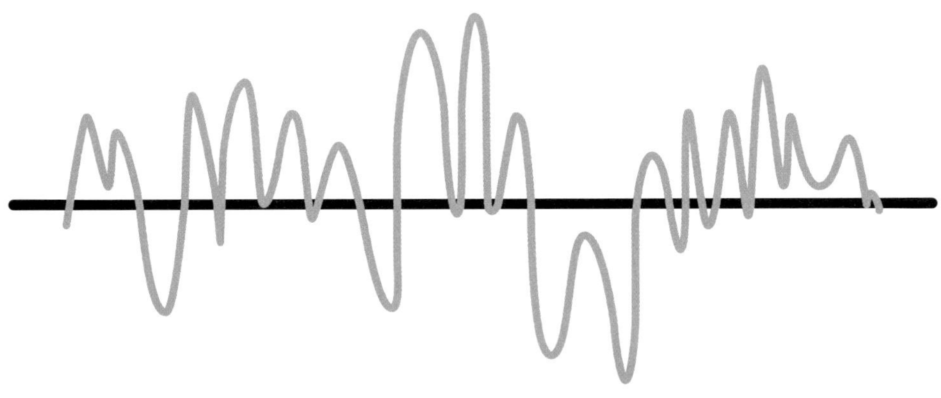

FIGURE 4-6: A *noise waveform* contains random spikes with no set pattern. It's used for producing drum sounds where pitch is hard to make out.

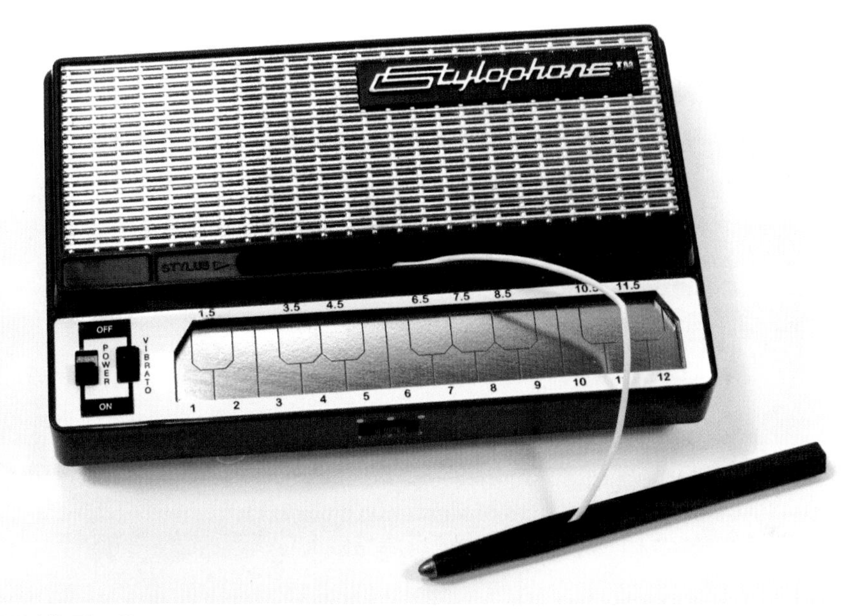

FIGURE 4-7: **The Stylophone, which uses touch pads and a stylus instead of regular piano-type keys, was one of the best-selling electronic instruments of the late 1960s and '70s. But when it was invented in 1968 by Brian Jarvis, it was considered a toy synthesizer. Its small size and unique buzzy (square wave) sound made it a cult hit with musicians like David Bowie. In 2007, the inventor's son released an updated version (shown here), which continues to be a favorite with musicians and fans.**

The Buzz about Synthesizers

A synthesizer is made up of several different electronic components that serve different functions. They can all be built into one unit, or you can connect separate modules together. The sound is created by generators, then shaped by modulators. Amplifiers make the sound louder or softer, and speakers turn the electric signals into sound waves that you can hear. Here are some common synthesizer components:

WHAT IS MIDI?

If you start to play around with electronic music, you may come across MIDI files. MIDI, which stands for Musical Instrument Digital Interface, is a computer code that lets you tell any kind of electronic musical instrument or device how to produce the sounds you are trying to create. It is the standard used by electronic music components and software to make sure that they all speak the same language.

***Oscillators* create oscillations or vibrations at different frequencies.**

They can be used in different ways. Audio oscillators are used as generators to create electrical signals with different frequencies. (*Audio* means the pitch of those frequencies is within the range of human hearing.) These produce the sounds that the synthesizer then shapes. Low-frequency oscillators can be used to control other waves. For instance, an oscillator can make the volume go up and down over and over, creating a wavering tremolo. And radio frequency oscillators create electrical signals that can be picked up by radios. (You'll play around with radio oscillators in the DIY Theremin project.)

A *filter* increases or decreases some of the harmonic frequencies of a tone.

Filters may affect only high frequencies, only low frequencies, or only frequencies within a certain range. This changes the timbre of the sound.

An *envelope generator* uses changes in voltage to control four stages of a note.

These stages are known as ADSR, which stands for attack (how the note builds at the start), decay (how the note fades at the end), sustain (how long the note is held), and release (how the note stops). The envelope can be applied to the loudness, the pitch, or the filter and make a big difference in the way a note sounds.

An *amplifier* makes the sound louder or softer.

You can control the amplifier by hand with a knob or automatically with an envelope generator.

MUSICAL INVENTORS: BALAM SOTO

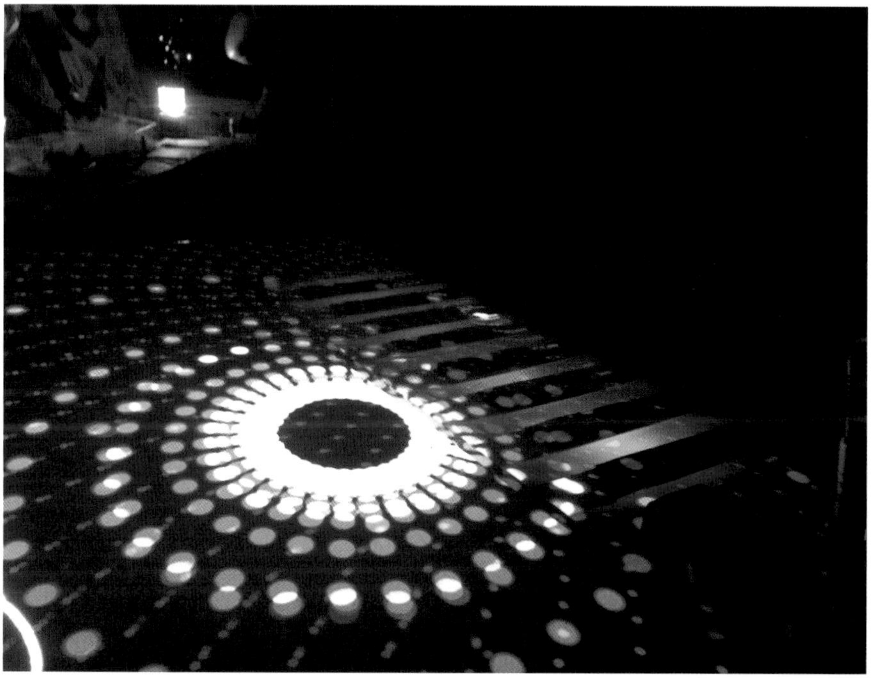

FIGURE 4-8: **Balam Soto's Exp.Inst.Textile is an interactive electronic art piece that lets you play notes by touching sensors on a cloth.**

Balam Soto is a new media artist who comes from Guatemala and now lives in Connecticut. Although he started out as a painter, his love of electronic music led him to create a series of experimental instruments that interact with the people playing them. His goal was to design something for anybody who wanted to make music but didn't know how to play an instrument. Some of his earlier models used clear plastic rods that looked like tall crystals and other controllers with lots of sensors that would respond to the player's touch. But for World Maker Faire New York in 2016, Soto wanted to make his instrument even simpler.

"I was thinking about making it friendlier. What is a material we are in contact with every day of our lives? Textiles!"

Once Soto decided to use textiles (another name for fabric or cloth) to build his instrument, he did some research on what makes an instrument easy to play. He found that a xylophone is considered the quickest instrument to learn. "You can figure it out in a matter of seconds."

So he built a "xylophone" out of cloth. The Exp.Inst.Textile is a flat piece of fabric that has seven sensors that respond to touch. The heart of the machine is an Arduino microcontroller board, a kind of mini-computer about the size of a credit card that can be programmed to read data from sensors and activate lights, speakers, and other devices. Using a computer program that he wrote, the Arduino takes in information from the sensors and translates it into notes. At the same time, the instrument projects colorful lights that dance over and around your hands in time with the music you are playing. Although the instrument is easy to play, Soto said building it involved many hours of hard work. "I sample the project step by step and test it. When I have problems, I have to undo all the previous steps and do the whole process all over again."

However, Soto believes that you don't need to start with special skills or training to become a Maker.

"Anything I need to know, I've learned myself. That's what I love about the Maker Movement. Anything we want to make, we're capable of learning ourselves."

The Strange and Mysterious Theremin

The theremin is one of the strangest kinds of electronic instruments. It consists of a box with two antennas. A loop-shaped antenna that comes out of the side controls the volume. The other antenna is a stick that stands straight up and controls the pitch.

Playing it is even stranger. Once you turn the instrument on, you never touch it. Instead, you control the sound by waving your hands closer and farther away from the antennas, opening and closing your fists. Shaking your hand makes the pitch wobble up and down. Watching someone play the theremin is like watching some kind of upper-body dance or martial art exercise, full of fluttering and graceful movements.

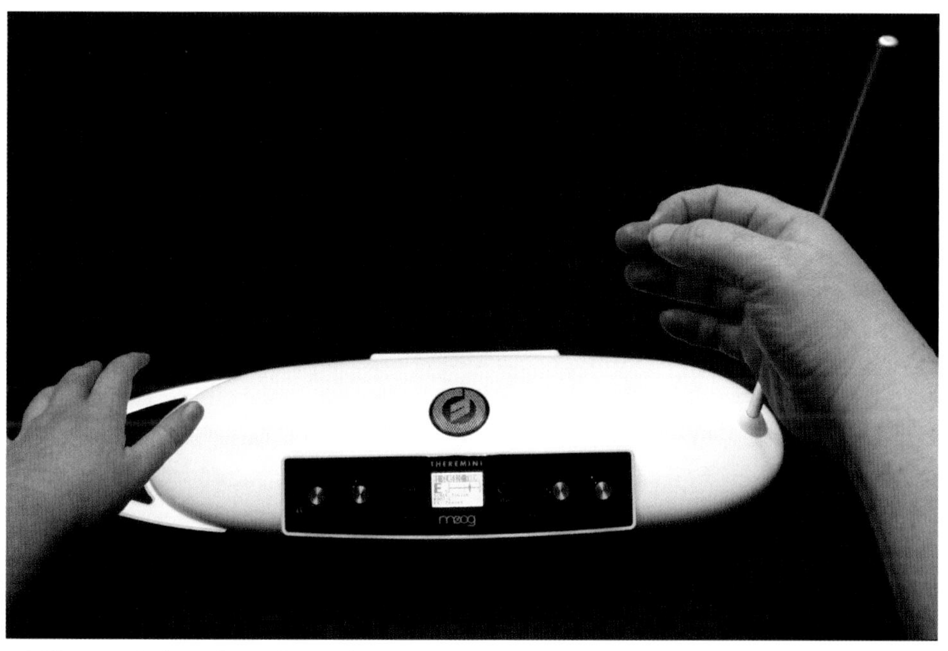

FIGURE 4-9: A modern Theramini by Moog. Just like with the original theremin, raising and lowering your hand above the loop on the left controls the volume. Changing the shape of your fingers near the antenna on the right controls the pitch. The modern version is part synth, with controls that let you adjust the scale and the waveform shape.

MUSICAL INVENTORS: LEON THEREMIN

The story of the man who invented the theremin is as strange as the instrument itself. Leon Theremin was a Russian scientist, who developed the instrument named after him in 1919. In 1927, Theremin came to New York City to promote his invention and teach other musicians to play it. His greatest pupil was a Russian violinist named Clara Rockmore, who became famous as a classical thereminist. In 1938, Theremin suddenly returned to Russia—his wife said he had been kidnapped by Russian spies—and was sent to work in a secret government laboratory. During his time as a prisoner, he developed a microwave-powered listening device that was hidden in a wall hanging and used for spying on American diplomats for seven years before it was discovered. After his release, he stayed in Russia and went back to building theremins and designing new electronic instruments.

The theremin has a unique and strange sound, too. Very few people are able to master the techniques needed to play it as a classical instrument, as it was designed. You're more likely to have heard it creating an eerie mood in old science fiction movies. Its spooky whistles and squawks sound a lot like the human (or alien) singers. It has also been used in rock music. That high-pitched voice singing "ooh-OOH-ooh, ooooh ooh" in the Beach Boys song "Good Vibrations" is actually a theremin!

If you've ever walked near a radio while it was playing and noticed a change in the sound—it gets louder, softer, or even squeals—you've experienced what happens when you play a theremin. A theremin works by creating interference between two radio waves. When the two waves mix, they combine and produce a new frequency, which you hear as a note. The pitch of the note is the difference between the two frequencies. Your body has its own, very weak, electrical field. So the closer you get to a device that's producing electrical waves, the more your electrical field affects those waves. That's how you can control the pitch of a theremin without even touching it!

Project: DIY Theremin

Materials

- 3 battery-powered AM or (AM/FM) radios, the smaller the better (Important: At least one of the radios must use a rotating dial as a tuner and be easy to open and close without damaging the components inside. This will be your variable transmitter.)
- Copper foil tape, aluminum foil tape, or thin, flexible wire
- Small telescoping antenna or stiff wire about 1 foot (30 cm) long, such as a piece of wire hanger or a bicycle wheel spoke
- Base for the antenna/wire, such as piece of cardboard or an index card
- Holder for the antenna/wire, such as a cork and glue, or a lump of clay

Tools

- Screwdriver (may be needed to open up the radio)
- Wire cutters (if needed to make the antenna)

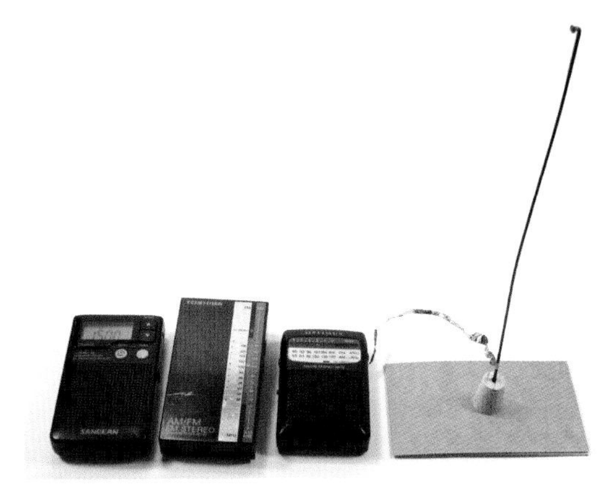

FIGURE 4-10: **You only need three ordinary portable AM radios to make a home version of a theremin.**

A theremin uses oscillators to create electrical waves, but you can make a simple version using ordinary radios. This project was adapted from instructions by sound artist and musical inventor Yuri Suzuki and Phil Borell. Old-fashioned radios work best for this project, so scour your attic, garage sales, or thrift shops for models from 20 or 30 years ago. See the Materials list for details on what to look for.

Safety Warning: Only use battery-powered radios for this project—do not use a radio that is plugged into your building's electrical supply! Radios that run on three or fewer AA or AAA batteries are safest.

Note: The foil tape or wire used in this project needs to be conductive. Copper tape with conductive glue is recommended. (For where to find it, see "The Musical Inventions Supply Closet" at the beginning of the book.) If the foil tape you are using does not have conductive glue, or if you're not sure, fold under a tiny bit of the tape at the ends. That way the shiny, conductive side will be touching the terminal to make a better electrical connection.

1. Before you make any changes to the radios, see how they affect each other just by tuning them to specific frequencies. When you tune in a station, you are actually adjusting the radio to pick up a certain radio wave frequency. The frequency is measured in kilohertz (written as kHz) for AM stations or megahertz (MHz) for FM stations. (Now you know why radio stations often use numbers as part of their name such as "FLY 92" or "KISS 106"—the numbers tell you their frequency!) The DIY Theremin uses the radios like this:

   - The first radio is a receiver. This is the only radio that you will hear.
   - The second radio is a fixed transmitter. Once you get it tuned to the proper frequency, it will cancel out the sound coming out of the receiver.
   - The third radio is the variable transmitter. By moving the frequency on this radio up and down with your body, you will change the pitch of the sound coming out of the receiver. Right now you will use this radio as is, but later on you will add an antenna (or use its antenna if it has one) so you can control it just like on a real theremin. To do this, the radio you use for a variable transmitter should use a rotating dial as a tuner and be easy to open up (by removing screws or by carefully prying it apart).

2. Here's what to do with the three radios:

- Take the radio you are using as a receiver, turn it on, and set the tuner to 1500 (which tells it to pick up a frequency of 1500 kHz). Turn up the volume. If you can hear talking or music, try going up the dial (raising the frequency) until you hear mostly static, that scratchy noise between stations.
- Turn on the second radio, the one that will be your fixed transmitter, and lower the volume until you can't hear it at all. Set the tuner to a frequency that is 455 kHz below the station on the receiver radio. In other words, if your receiver is set at 1500, set the fixed transmitter at about 1045 (1500 – 455 = 1045). If the receiver is set at 1600, the fixed transmitter should be set at 1145. Hold the fixed transmitter right next to (or even on top of) the receiver. The static should disappear and the receiver should go silent. If it doesn't, try adjusting the frequency on the fixed transmitter until you get as little sound as possible from the receiver.

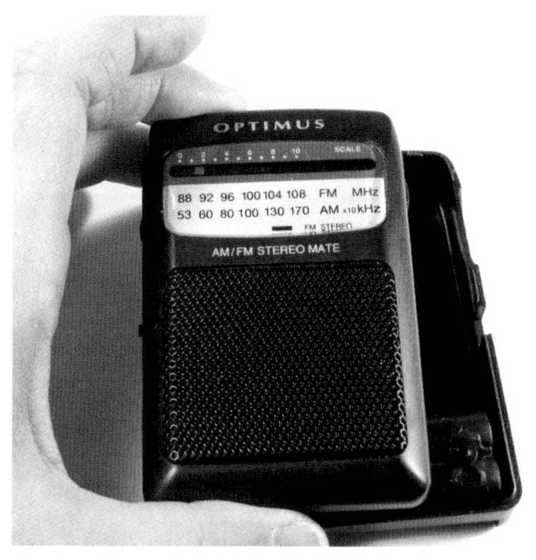

FIGURE 4-11: The variable transmitter radio should have a tuning dial and be easy to open.

FIGURE 4-12: Inside this radio is a circuit board that covers the back of the tuning box (marked with a square). Be careful not to break the wires connecting the back and front together.

- Turn on the third radio, the variable transmitter, and lower the volume until you can't hear it at all. Set the tuner to the same frequency as the second (fixed transmitter) radio. Hold the variable transmitter radio near the first two radios. Instead of silence, you should hear a whistling sound that changes as you move it closer and farther away. Put the variable transmitter radio down near the other radios and see how moving your hand around it affects the pitch. Then try changing the frequency with the tuner slightly to change the pitch. If you want, you can stop here and use your theremin as is. But if you want to add an antenna, go on to the next step.

3. To alter the variable transmitter so you can play your DIY theremin using an antenna, first tune it to an AM station until you can hear talking or music. Then, carefully open up the radio. You may have to remove some screws (which may be hidden underneath the batteries) or carefully pry the back off. Avoid breaking any wires, especially any that connect the front and the back.

4. Look for the plastic box that surrounds the tuning dial. On the back of the box (or on the circuit board it is attached to), near the corners, there should be four screws or blobs of silver solder (the melted metal used to connect components to circuit boards). One of these is the capacitor terminal, the component that tunes the radio to AM stations. Touch each blob with a small screwdriver until you find the one that makes the station more staticy and harder to hear. When you find the capacitor terminal, stick one end of a piece of foil tape or wire to it. The tape should be a few inches long, but thin

FIGURE 4-13: **The round tuner dial (black disk with ridges around the edges) is connected to the tuner box (white-ish box). The circles show the shiny silver blobs of solder that connect the box to the circuit board. One of these is connected to the tuning for the AM frequencies.**

FIGURE 4-14: **Stick the foil tape onto the blob that changes the tuning.**

enough to avoid touching anything else besides the terminal. Carefully close the radio back up.

5. Make an antenna and a stand that will hold it upright. A piece of stiff wire stuck into a cork and attached to a piece of cardboard is fine. Attach the other end of the foil tape to the antenna. If there already is an antenna built in, attach the other end of the tape to that.

FIGURE 4-15: **Close the radio back up, being careful not to break the foil tape. Attach the other end of the tape to the antenna.**

6. Stand the homemade antenna next to the variable transmitter radio. Tune it to the original frequency (the same as the second, fixed transmitter, radio). Use the tuning dial to adjust it until you hear the whistling tone again. Try to get the lowest pitch you can. Then slowly wave your hands around the antenna to see how many notes you can produce. Wiggling your hand creates a wobbly vibrato. What other effects can you get out of your new theremin?

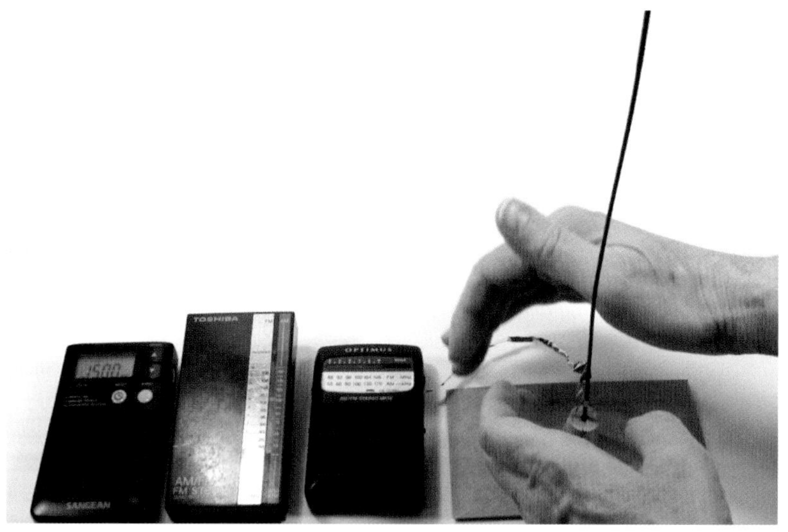

FIGURE 4-16: **To play the DIY theremin, move your hands around to produce different pitches. You may have to wave them over the radios to achieve some notes.**

PLAYING AROUND WITH SYNTHESIZERS

Want to see how synthesizers work for yourself? If you don't have access to an actual synth, try one of these simulations:

- **Moog Synthesizer Google Doodle (google.com/doodles/robert-moogs-78th-birthday):** This free online simulation lets you twirl knobs and push buttons to produce different sounds.
- **Clara Rockmore Google Doodle (google.com/doodles/clara-rockmores-105th-birthday):** Take theremin lessons with the best player of all time, then fool around with an online virtual model with waveform and other controls.
- **Online Tone Generator (onlinetonegenerator.com):** Hear the difference in timbre between the different waveforms by typing in a frequency and listening to it in the form of a sine, square, sawtooth, and triangle.
- **Chrome Music Lab (musiclab.chromeexperiments.com):** A music education site from Google that lets you play around with sounds and animated images. "Experiments" on the page include fun cartoon keyboards, sound waves, and oscillators. You can also click on links to see the computer code used to build them.

Project: littleBits Synth Glove

Materials

- littleBits Synth Kit, including a 9V battery
- Additional littleBits modules:
  - Wireless Transmitter and Receiver
  - Light Sensor
- 3 hook-and-loop shoes
- 1 or 2 wire modules
- Short strip of peel-and-stick hook and loop tape (just the scratchy hook side)
- Old pair of gloves, the longer the better (fingerless gloves look especially cool!)
- Small area with a bright light (such as under a lamp or by a window)

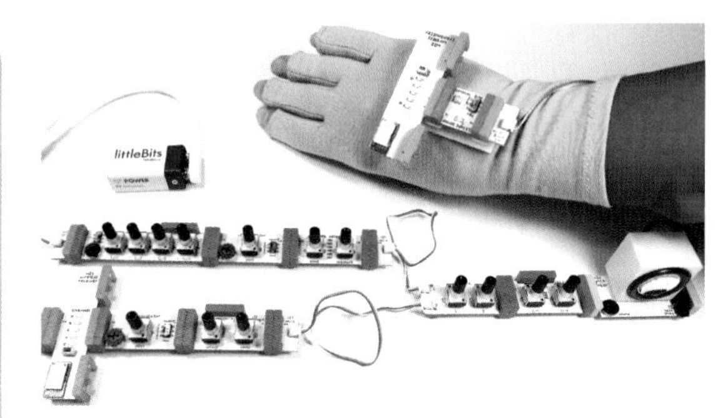

FIGURE 4-17: The littleBits Synth Glove lets you control the music you create with hand movements, even from across the room.

The littleBits Synth Kit is made up of small, snap-together magnetic parts that can be arranged in different combinations to create a real working music synthesizer. It's based on the Korg MS-20, a synthesizer that was first introduced in 1978. The Synth Kit lets you shape sound waves, control them with switches and sensors, and even program repeating patterns. And the kit is powered by a 9V (9 volt) battery, so it's safe to take apart and put together.

Note: The additional littleBits modules in the materials list are all found in the Gizmos and Gadgets kit, Version 1. The project can also be adapted to use the Bluetooth Low Energy bit in the second version of the kit. If you don't have either, connect the glove to the synth with wire modules.

If you don't have littleBits hook-and-loop shoes to attach the modules to the glove, you can try peel-and-stick hook-and-loop tape or stickers, found in crafting and hardware stores.

The stretchy fingerless gloves used here are Dritz Creative Comfort Crafter's Comfort Gloves.

FIGURE 4-18: A Micro Sequencer module from the littleBits Synth Kit

All the elements of a basic synthesizer are found in the kit:

- Signal generators (two oscillators and a random voltage generator)
- Controllers (a keyboard and a sequencer)
- Modulators (an envelope that controls how quickly each note gets louder and softer, and a filter that lets you adjust the balance of high frequencies and low frequencies)
- Modifiers (a delay that controls how much echo a note produces, and a mix module that combines sounds from two sources and sends them to the speaker)

You can connect the modules in multiple ways to create an endless variety of sounds. And each of the modules has knobs, dials, and switches that let you adjust the settings to your taste. Best of all, you can add other kinds of littleBits modules and attach them temporarily to creations you build using crafts materials or even your clothing to make your own synth-based musical inventions!

The littleBits Synth Glove makes it possible to play different notes and phrases with a wave of your hand. It was inspired by the Musical Glove invented by Imogen Heap (see the "Musical Inventors: Imogen Heap" sidebar that follows this activity).

1. First build the synth circuit. The circuit has two rows. The top row uses the Micro Sequencer to play a repeating series of four notes as background. The bottom row creates a variable tone that you will control with your glove. Both rows start with an oscillator to generate the sound wave frequency. Then they add other modules to get a more interesting

sound. Finally, they are blended together with the Mix module and sent to the speaker, which converts the electrical signal to a sound wave. (See Chapter 5 to find out how speakers work!) Start by building the top row using the following modules:

Power > Split > Micro Sequencer > Oscillator > Delay > Mix > Filter > Synth Speaker

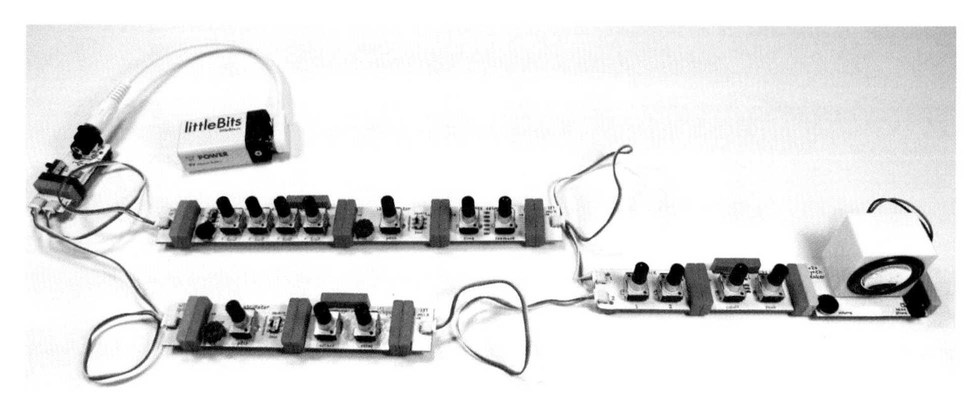

FIGURE 4-19: **The whole circuit for the littleBits Synth Glove includes two rows with oscillators, which are connected with a mixer that leads to a speaker.**

FIGURE 4-20: **The top row of the circuit**

FIGURE 4-21: **The speaker part of the circuit**

Next, build the bottom row. Take the extra wire on the Split module and add the following modules. At the end, connect the row to the extra wire on the Mix module.

(Power > Split) > Oscillator > Envelope > (Mix > Filter > SynthSpeaker)

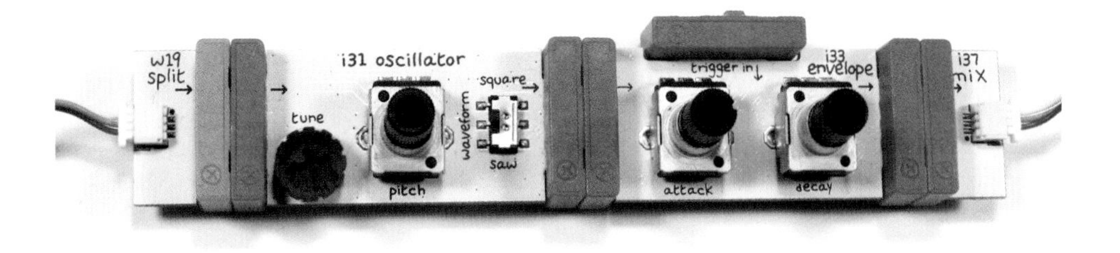

FIGURE 4-22: The bottom row of the circuit

> **Note:** For this project, you can use any arrangement of modules from the Synth Kit that is controlled by the amount of voltage going into the circuit instead of a keyboard. Find other ideas in the instruction guide that comes with the kit or online at *littleBits.cc/synth*. Look for circuits built around a Random module, then substitute the Light Sensor module for the Random module to create variable voltages.

2. Now you can create some music, beginning with the background sound. To create the background rhythm that will repeat over and over, set the tiny switch on the Micro Sequencer to "clock." Tune the Sequencer to play different notes by adjusting the four knobs. Choose the speed with the little dial. To change the timbre, play around with the Delay module, or swap it out for another one. The other row of modules produces the variable note. Tune it to a note that sounds good with the background rhythm. This is the tone that will be controlled by the glove.
3. When you like the way your synthesizer sounds, it's time to add the main layer of sound. Adjust the modules on the second row using all the knobs, dials, and switches until you get a note that sounds good with the background rhythm.

4. When you're happy with both layers of sound, insert the Wireless Receiver module between the Split wire and the Oscillator. Make sure it is set to the same channel as the Transmitter. Use the middle connector on the right side (number 2). You can turn off this circuit while you build the glove controller.

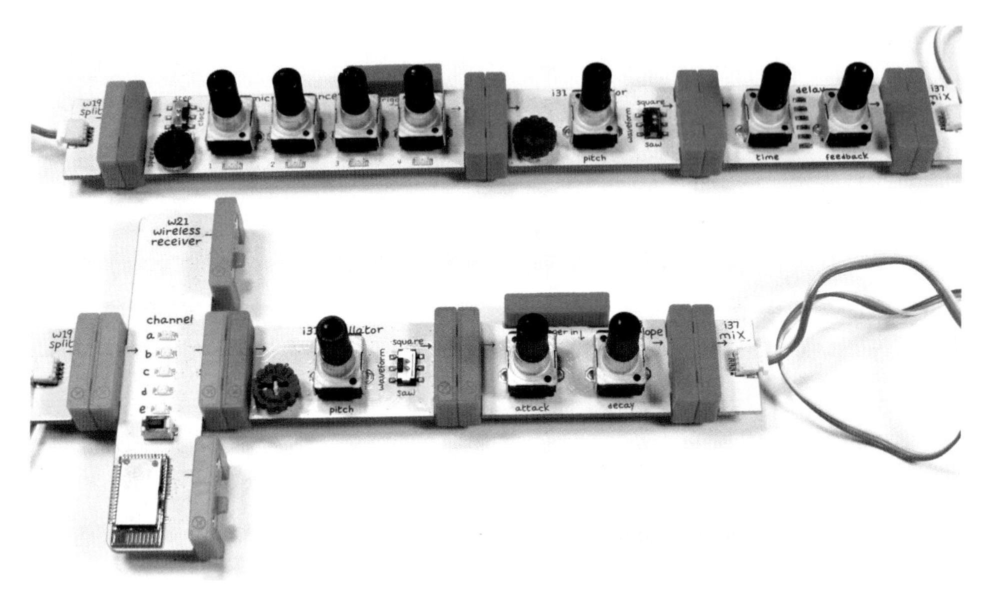

FIGURE 4-23: Insert the Wireless Receiver in the bottom row of the circuit, right before the Oscillator module.

5. The glove controller lets you play your synth by moving your hand around. It varies the voltage according to the amount of light shining on it, and it will work even if you are standing on the other side of the room from the synth, up to 100 feet (30 m) away! Make sure the Wireless Transmitter is set to the same channel as the Wireless Receiver and use

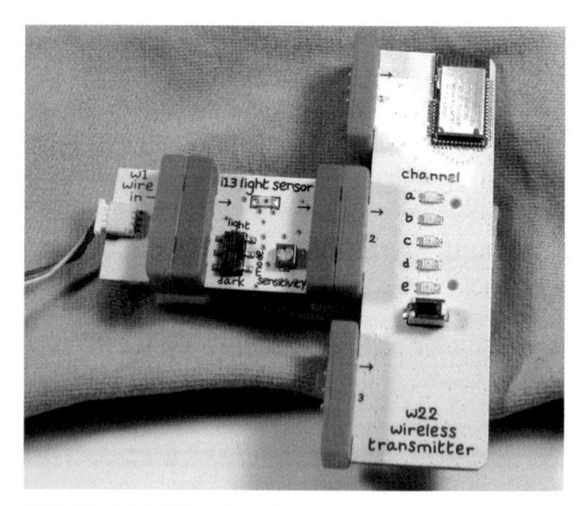

FIGURE 4-24: The glove circuit

the middle connector on the left side (number 2). The circuit should look like this:

Power > Wire (you may need two, attached end to end) > Light Sensor > Wireless Transmitter

6. To attach the Light Sensor circuit to the glove, you'll need to use the hook-and-loop shoes. They also help keep the modules connected to each other. Snap the shoes over the feet where the modules meet. Snap the third shoe onto the other foot on the Wireless Transmitter. Cut a piece of hook-and-loop tape long enough to fit all three shoes. Press the pieces onto the hook-and-loop strip until they are firmly attached, leaving the paper on. To position the tape, put on the glove and make a fist. Peel off the backing paper and stick the tape onto the back of the glove so the top of the T shape goes across your knuckles. Press it firmly in place.

FIGURE 4-25: Connect the bits in the glove circuit by snapping littleBits hook-and-loop shoes over the feet on the underside of the modules.

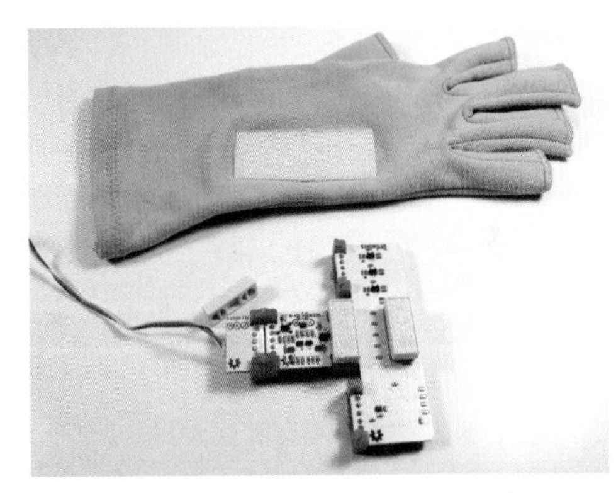

FIGURE 4-26: The glove circuit next to the glove, showing where the hook-and-lock tape will go

7. Now play your musical glove! Turn on your synth and let the rhythm start going. Turn the glove circuit on, and slip the battery into a pocket, where the wires won't get in your way. Then stand in a spot where you can wave your hand in and out of the light. How does the sound change? Can you control the pitch, the volume, or the timbre?

Troubleshooting tips:

- If either of your circuits isn't working, check to see that all the modules are connected properly. The Wireless Transmitter and Receiver have on-board LED indicator lights that show you when they are getting power from the battery.
- If you don't get a lot of variation in sound as you move your glove around, use the little plastic screwdriver that comes with the sensor to adjust the sensitivity screw.

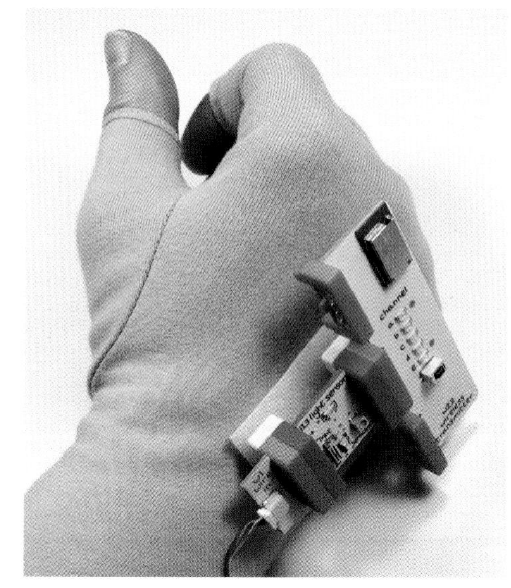

FIGURE 4-27: Stick the hook-and-loop tape to the glove so the Wireless Transmitter sits along the line of your knuckles.

- If there's not a lot of light around, try switching the sensor to "dark" instead of "light" and let it measure different degrees of darkness.
- The Light Sensor may work better if you use an LED light source or direct sunlight coming in through a window.
- Use fresh 9V batteries—the Wireless modules seem to burn them out fast.

Extensions and adaptations:

- Try adding different kinds of sensors instead the Light Sensor. How can you make the sound of the synth change by bending a finger or clapping your hands?
- Find tips on how to record, edit, and share your music at *littleBits.cc/recordyourmusic*.

MUSICAL INVENTORS: IMOGEN HEAP

In 2012, musician Imogen Heap gave the first demonstration of what she called her "magical musical gloves." The gloves contain sensors and Bluetooth technology that let Heap control the music-making software on her computer while she is moving around stage. Instead of turning dials or pushing buttons, Heap can play notes or change the sound of her voice by making different gestures—opening and closing her hand, swinging her arms, making hitting motions—and by walking around to different areas.

To turn her dream of musical gloves into a reality, Heap gathered a team of people who knew how to combine fashion and technology. They designed clothing where the wiring also served as decorative stitching and the hardware and circuit boards were built into a frilly vest. Heap's group formed an organization called mi.mu to develop the gloves and encourage others to offer their own ideas and suggestions.

Other musicians, including Ariana Grande, have experimented with Heap's musical gloves and have even used them in performances. Mi.mu is continuing to work on the gloves, with the hope of one day making them available to artists of all kinds.

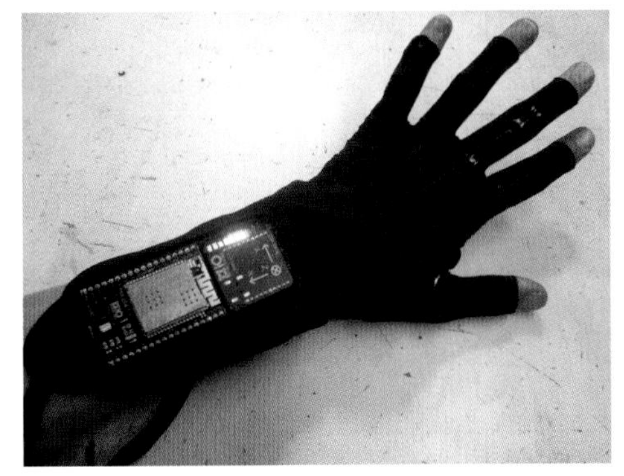

FIGURE 4-28: A recent version of the mi.mu musical gloves invented by Imogen Heap

Project: Makey Makey Musical Surface

Materials

Large sheet of heavy paper or thin cardboard

Aluminum foil tape (or other conductive sheet, such as aluminum foil, and a glue stick)

Tools

Scissors

Laptop or desktop computer (must have a keyboard) and Internet access or software for creating music (such as Scratch)

Small remote control (RC) car and controller

Makey Makey kit or Makey Makey GO

(Optional) extra-long alligator clip wire (also called a "test lead")

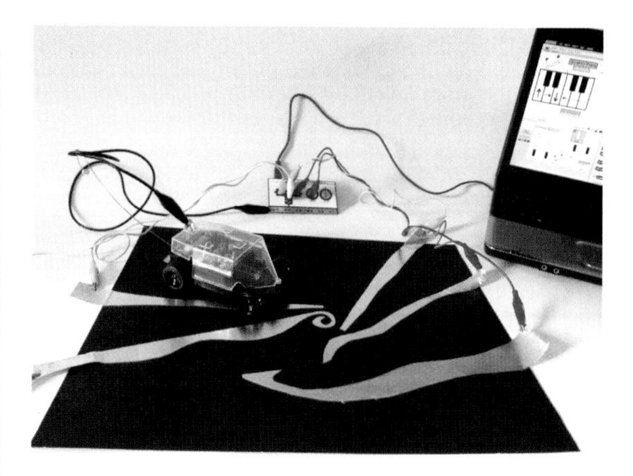

FIGURE 4-29: **With a Makey Makey, you can make an electronic instrument out of cardboard, aluminum foil, and an RC car.**

One of the fun things about electronic instruments is that they don't have to look anything like "normal" instruments. They don't even have to look like the control board of a synthesizer, with knobs and buttons. In fact, you can use anything, including drawings or other kinds of art, to trigger electronic sounds. All you need is something that can control an electrical circuit to produce a sound. That means you can turn ordinary arts and crafts into music synthesizers!

One easy way to do that is with a Makey Makey invention kit. Makey Makey was created by inventors Jay Silver and Eric Rosenbaum when they were students at the MIT Media Lab. It is an Arduino-based microcontroller board that plugs into your computer and acts like a second keyboard. (There's also a smaller version called Makey Makey GO that has one key.) The Makey Makey board has six metal pads marked up, down, left, right, space, and click on the front, and places on the back where you can plug in wires to create even more keys. When you connect something conductive to a key on the Makey Makey, it triggers that key—just as if you had pressed it on your computer keyboard. When you use it with Scratch (a free online programming language for kids created by MIT) or with other music software, it lets you turn unusual objects into a synthesizer.

This project shows you how to use Makey Makey and a remote control car to create a musical playing surface. Try it out, and then let it inspire you to invent a musical controller of your own!

1. First, make the playing surface. Cut the aluminum foil tape into strips, one for each key that you want to connect to a note. You can trim the tape into interesting shapes if you like. Then fold over one end, with the protective paper on the inside. This will be the tab that hangs off the board so you can clip the Makey Makey's wires onto the conductive areas. Peel back the protective paper on the folded end of the tape to expose the glued side. Press it so the glued sides stick to each other. If you need to, trim off any of the glue that is sticking out.

FIGURE 4-30: Cut the peel-and-stick aluminum foil tape into strips or different shapes.

2. Attach the first strip anywhere along the edge of the board with the tab hanging off. Slowly peel off the rest of the protective backing while smoothing the tape down on the cardboard. Repeat with the other strips—making sure that none of them are touching. You

FIGURE 4-31: To make a tab to clip a wire to, peel off some of the foil tape and fold it over so the glue sticks to itself.

may also want to test the RC car to make sure it can drive around the board without getting stuck.

FIGURE 4-32: **Attach the rest of the foil strip to the cardboard, with the tab hanging off the end.**

3. Now connect the Makey Makey to the computer by plugging the cord that comes in the kit into a USB port. You do not need any other software or drivers. The lights on the board should go on. To test the "keys" on the front of the Makey Makey, touch the metal Earth bar along the bottom edge of the board with one hand, and touch one of the metal pads for the keys with the other. You should see the light next to that key light up. If it doesn't, check the troubleshooting tips at the end of this activity. To see how the Makey Makey triggers a note, go to *www.makeymakey.com*, click on the Apps page, and try the piano or one of the other sample music apps. The Makey Makey will work with any program that uses the keys on your computer keyboard.

4. Next, test the playing surface to make sure it works with the Makey Makey. Clip one of the alligator clip wires that come with the Makey Makey into one of the keys on the board. The jaws of the clip fit into the holes on the board. Clip the

FIGURE 4-33: **This playing surface has five notes.**

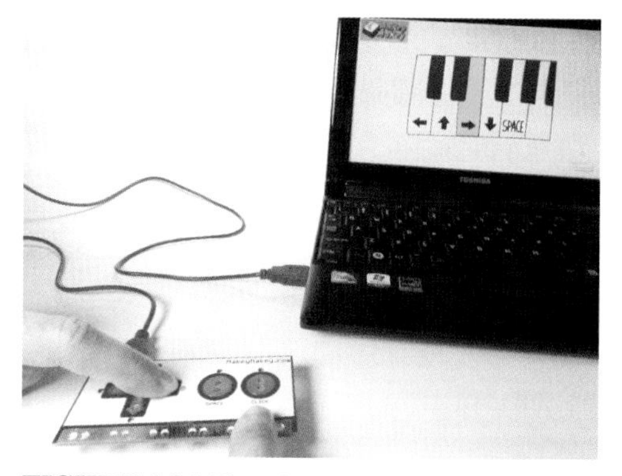

FIGURE 4-34: **Test the Makey Makey by touching the Earth bar and one of the keys at the same time. You should see a light near the key go on. The note triggered by that key, in this case the Right Arrow, should light up and play.**

other end of the wire onto the tab for one of the strips. Touch the Earth bar and one of the foil strips. The key should trigger, just like when you touched the board itself. Check the troubleshooting tips at the end of this activity if you need help. Do the same with all the other foil strips on the playing surface.

5. It's time to connect the RC car. Fold under the end of a new strip of foil tape. Attach it to the back of the car to make a tab that drags along the ground a little. The bottom tab needs to touch the conductive strips on the playing surface without getting stuck or coming off. The foil strip should be long enough to fold into another tab at the other end that sticks up. The top tab needs to keep the wire you will connect to the car away from the playing surface so it doesn't get tangled. Connect a wire (extra long if you have it) to the top tab. Connect the other end of the wire to the Earth bar. If your wire isn't long enough, connect two alligator clip wires together by snapping the metal teeth of one over the metal teeth of the other.

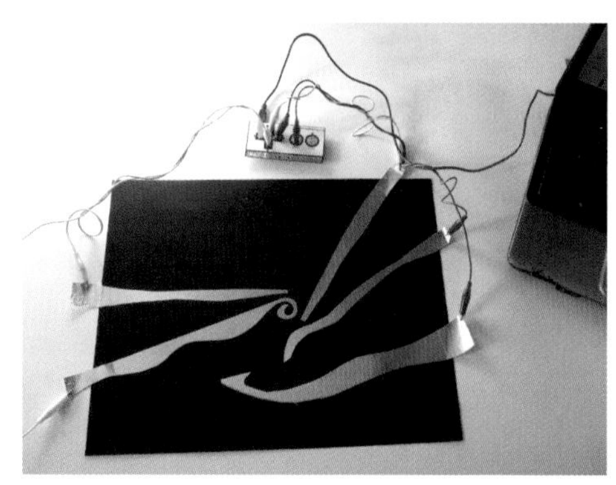

FIGURE 4-35: Use the alligator clip wires to connect all the foil "keys" to the Makey Makey.

FIGURE 4-36: An RC car and its remote

FIGURE 4-37: Attach a strip of foil tape to the back of the car. The strip should be long enough to fold under to make tabs at the top and the bottom.

6. Now that your musical surface is working, you need something to play! The Apps page on the Makey Makey website has links to online instruments you can try. (Some of them are listed in the following "Music Software for Your Makey Makey" sidebar.) You can also use any computer program or website that uses the arrow keys to make sounds. And to really have fun, check out programming languages like Scratch and other software that lets you write your own music. Here are some ideas for how to use them:

   - Use the keys to play individual notes in a scale or the notes of a song.
   - Use the keys to play a phrase made up of multiple notes.
   - Record your own samples or upload music or sounds and shape them using the music software.
   - Use one or more of the keys to shape the sound of the other keys, such as by changing the timbre or the pitch. For example, on Scratch, you can play notes that sound like different instruments by plugging

FIGURE 4-38: The bottom foil tab should drag along the ground, where it will rub across the foil on the Playable Surface.

FIGURE 4-39: The top tab should stick up to give you a place to connect the alligator clip wire where it won't get tangled as you drive the RC car around.

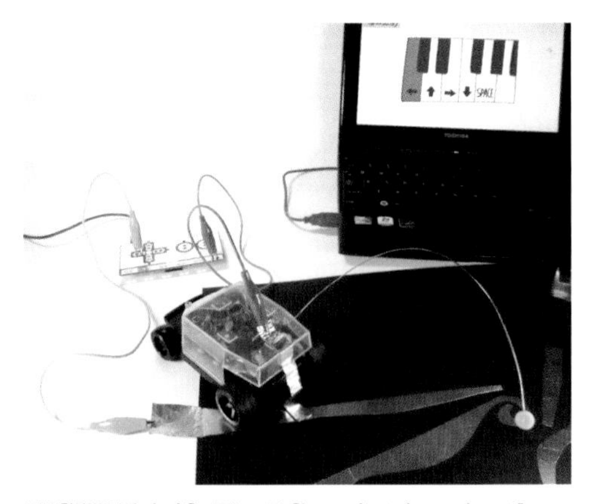

FIGURE 4-40: The RC car is triggering the piano key shown on the computer screen in red.

the number of the instrument into the program. If you write a Scratch script (program) that tells it to add 1 to the number each time, you'll get a different instrument every time you hit that key.

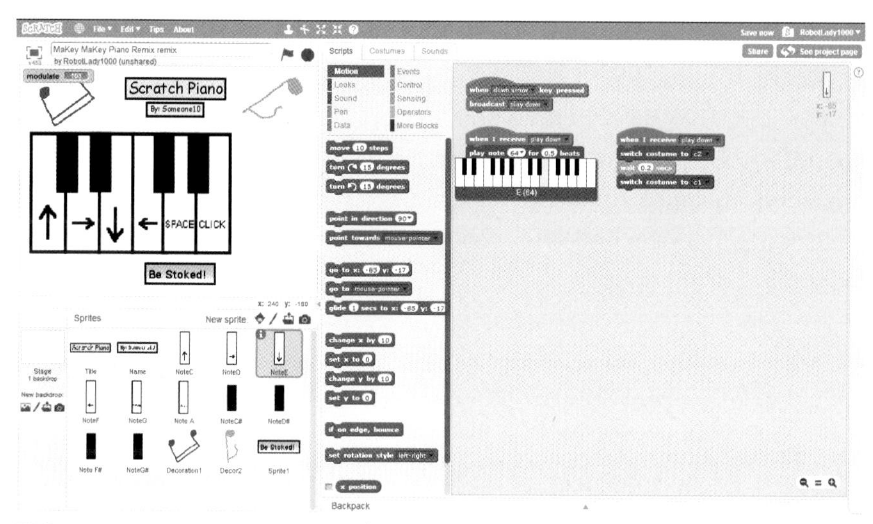

FIGURE 4-41: **The Scratch Piano (*scratch.mit.edu/projects/2543877*) can be played with the arrow keys. Click on the windows with the numbers inside the drag-and-drop commands to change the notes it plays. You can even change the keys to sound like different instruments!**

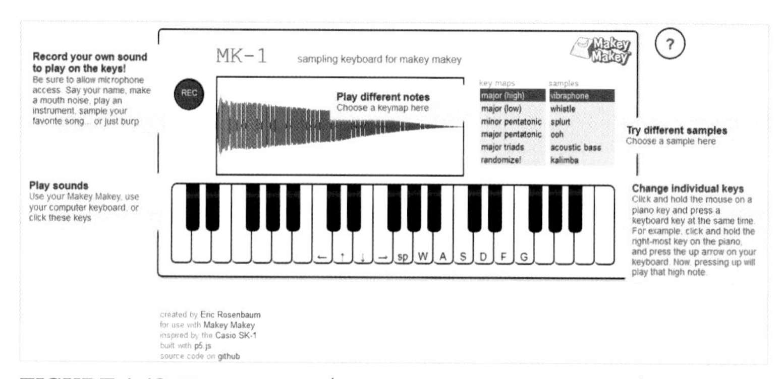

FIGURE 4-42: **The MK-1 app (*ericrosenbaum.github.io/MK-1*) by Makey Makey cocreator Eric Rosenbaum**

Troubleshooting tips:

- Don't forget that you must connect to both a key and the Earth bar to activate a note.
- If you don't get a sound when you touch the Makey Makey keys or foil strips on the musical surface with your fingers, your skin may be too dry. Use hand cream to moisten it.
- If two notes sound at the same time when only one is triggered by the RC car or your finger, then there is a short circuit connecting them. Look for any place the foil strips may be touching. There should be space around every strip.

Adaptations:

- To create a musical playing surface with a Makey Makey GO (which only has one key), use Scratch or other music software to write a program that plays a song segment by segment, or that generates random notes every time it is triggered.
- If you don't have an RC car, you can play the surface with your hands by attaching the grounding wire to the cuff of your shirt or a bracelet. Make sure that it touches your skin. Or make a pointer out of conductive material (such as aluminum foil) that you can use to activate the keys.

Extensions:

- Hide the tangle of wires connecting the surface to the Makey Makey by building it into or on top of a cardboard box base.
- To get rid of the Earth wire attached to the RC car, put a strip attached to Earth next to (but not touching) every one of the strips that activate the keys. Then the strip attached to the car (or even foil wrapped around the wheels) will act like a bridge to connect the two strips and activate the key.
- Experiment with other ways to make a musical playing surface. What about a bumpy design made with conductive modeling dough? (Don't let the lumps of dough touch, or separate them with non-conductive Plasticine clay.)
- Decorate the RC car, or build a cover for it to go with the theme of your playing surface. For instance, you can make it look like a hand and have

it drive around a surface that looks like a piano keyboard. Or go wild and turn it into a cloud that floats across a surface that looks like the sky, or a mermaid that "swims" around an underwater scene.

MUSIC SOFTWARE FOR YOUR MAKEY MAKEY

Here is a list of some of the software you can use with your Makey Makey Musical Surface project. Most are included on the Apps page of the Makey Makey website (*makeymakey.com/apps/*).

- **Scratch Piano (*scratch.mit.edu/projects/ericr/2543877*):** Play the piano with the arrow keys, then remix it using the free online (or downloadable) Scratch programming language from MIT.
- **MK-1 (*ericrosenbaum.github.io/MK-1*):** A sampling synth for Makey Makey. Record your own sound, use preset scales, and set the keys to play just the notes you want.
- **Makey Makey Harp (*scratch.mit.edu/projects/59631970*):** This Scratch program plays random notes using only one key. It was designed to work with Makey Makey GO.
- **Chamber Music Piano (*www.nyu.edu/projects/ruthmann/CMSD/piano*):** Play along with a piano video using the Makey Makey keys. The notes are all on the pentatonic scale, so they always sound good!
- **Soundplant (*soundplant.org*):** Download this program to your computer and you can drag and drop sounds onto your typing keyboard. You can try a trial version for free.

The Silly Science of Circuit Bending

Circuit bending is a goofy way to create musical sounds. You take an electronic device—usually a singing doll or a toy that talks—and rewire the insides to change the way it sounds. The random squeaks and noises can then be used to create electronic music. Taking a prerecorded sound and distorting it to create something new is called *sampling*. In a synthesizer, samples are used just like other kinds of sounds. You can modify a sample by playing it faster or slower, which makes the pitch go higher or lower. You can also trim it into shorter pieces, loop it around over and over, or play it in reverse. Some musicians even play circuit-bent instruments live in concert.

FIGURE 4-43: **This Speak & Read "Fury Incantor" created by the inventor of circuit bending, Reed Ghazala, uses electrified parts of a classic car called the Plymouth Fury as body contacts that you control by touching them.**

Musical Inventor Reed Ghazala (*anti-theory.com*) is known as the "father of circuit bending." As a teenager in 1967, he accidentally short-circuited the amplifier from a toy by touching it to his metal desk and it started to squeal. He began adding knobs and switches to turn toys into toy-shaped music synthesizers, and his creations became prized by top rock musicians. That started a musical maker movement that is still popular today. However, modern talking or musical devices use miniature components on printed circuit boards, which are harder to rewire. So circuit benders are always on the lookout for old toys that still have components like resistors they can remove and replace to create their own inventions.

MUSICAL INVENTORS: DAVE BARNES

FIGURE 4-44: The toy guitar I found in a thrift shop for $2 turned out to be a Dr. Freakenspeak classic. On the back was a label with a serial number and contact information for its maker, Dave Barnes. The metal pads and the knob at the bottom are his circuit bending additions.

While combing thrift shops for toys that could be circuit bent, I happened across a toy guitar in a sale bin that had already been transformed—by a circuit bending master! Dave Barnes of Canada, also known as Dr. Freakenspeak, has turned toys into musical tools that can be used as sound sources for sampling or live performances. Well-known musicians like Sean Lennon, Joe Walsh, and Imogen Heap (who invented the Musical Glove mentioned earlier in this chapter) have used Freakenspeak circuit bent creations in their compositions.

"I love the process of discovering new bends on a toy or device that I have not bent before," Barnes says. "I am always surprised with the interesting sounds I can find. My other favorite part of bending is building the toy circuits into new and interesting enclosures like vintage camera cases."

Barnes said that the Circuit Bent Keytar I found was made around 2005. It has two basic bends: a pitch control knob, and two metal pads that serve as "body contacts" that you turn on and off simply by touching them, making your body part of the circuit.

For beginners who are interested in circuit bending, Barnes recommends looking for toys made in the 1970s and '80s. "My favorite units to bend are Furby (not the new one but the one from the 1980s), the Speak & Spell (including the math and reading versions), and the Touch & Tell from Texas Instruments. My best advice for anyone wanting to get into circuit bending is to get a copy of Reed Ghazala's book *Circuit-Bending: Build Your Own Alien Instruments*. It has lots of great info and is really the 'bible' of circuit bending."

To see demos of instruments created by Dave Barnes, check out the Freakenspeak YouTube channel (*youtube.com/user/FREAKENSPEAK*).

Project: Simple Circuit Bending

Materials

Musical greeting cards

Pencil, preferably with a soft lead, such as a 2B art pencil (the higher the number on a B pencil, the softer the lead)

Index card or small piece of thin cardboard or heavy paper

2 alligator clip wires (also known as test leads)

Paper and pen for taking notes

(Optional) solar panel (may be taken from a solar garden light)

(Optional) light sensor (See "The Musical Inventions Supply Closet" in the front of the book for sources.)

Tools

(Optional) hole puncher

(Optional) camera/video recorder/audio recorder (you may be able to use a cell phone)

(Optional) hot glue gun (or soft-drying flexible clay, such as Model Magic)

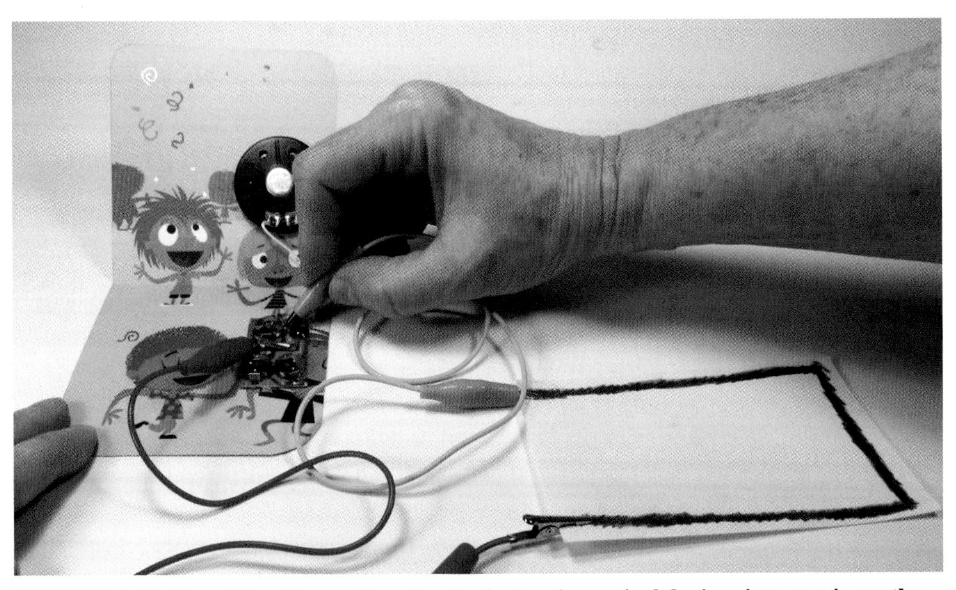

FIGURE 4-45: Circuit bending makes simple electronic musical devices into music synthesizers—even musical greeting cards!

There are some quick and easy ways to play around with circuit bending on modern circuit boards that let you create different kinds of sounds. It's harder to use these as synthesizers, because they won't make the exact same sounds over and over when you them want to. But you can record the sounds you produce and use them with sampling software. (See the Makey Makey Musical Surface project for software suggestions.) And they're fun to experiment with!

> **Safety Warning:** Circuit bending should only be done on battery-powered devices that use 9 volts or less! To avoid a dangerous shock, *NEVER* open up any electrical or electronic devices that plug into the wall. Circuit boards can contain poisonous materials, so do not touch them and then put your fingers in your mouth (the classic "lick and stick" method of circuit bending). Also avoid connecting parts of the circuit board directly using wires or other metal connectors, which can cause the device to overheat and possibly start a fire. See the instructions for a safer method!

1. Open up the greeting card. The prerecorded music should start to play. (You may want to make a video or audio recording of the music to help you remember what it sounded like before you started to "bend" it.)

FIGURE 4-46: The inside of a musical greeting before it has been hacked

2. The first thing you need to do is make your own switch so you can turn the music on and off while the card is open. Pry open the glued flap inside the card to expose the circuit board and speaker. Look for a little plastic tab that slides in and out of the circuit as you open and close the card. This tab separates a metal clip from a metallic pad attached directly to the card. The metal clip is attached to a battery. (There may be more than one battery on your card.) It acts like a drawbridge that lets electricity flow across it to make the sound play.

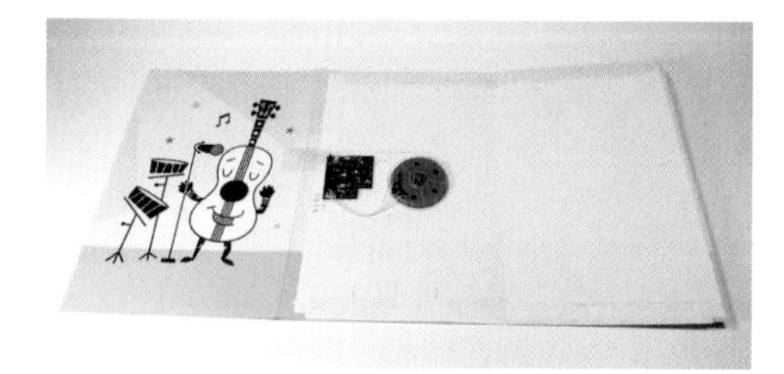

FIGURE 4-47: The card opened up to show the circuit and speaker attached to the inner layer

When the clip is up, the circuit is open and nothing happens. When it comes down and touches the pad, it closes the circuit and the sounds start to play. Remove the tab and replace it with a piece of paper, or cut off the tab with a piece of the card attached as a handle.

3. If you want, use the card (or another piece of stiff paper) to make a little stand. (To protect the wires coming out of the speakers, which can snap off if they are bent, cover them with a little hot glue or soft-drying flexible clay.) Fold the card so the speaker is standing up to make it easier to hear. Punch some holes on the paper behind it to let more of the sound out.

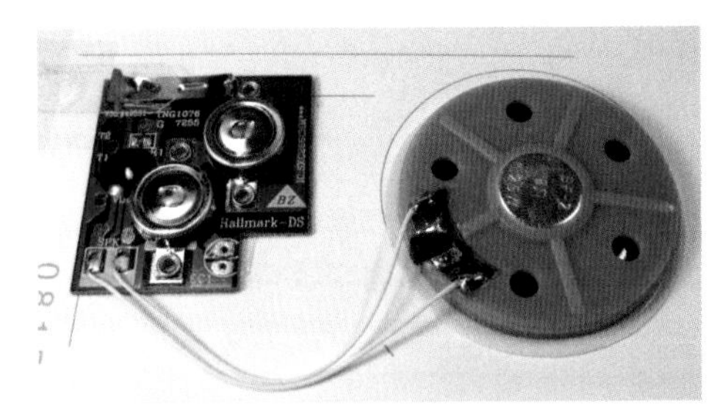

FIGURE 4-48: **A close-up of the circuit board and the speaker. Note the metal on/off clip at the top, the two round batteries, and the tiny metal components that can be used to bend the circuit. The component marked "R1" is a resistor and is a good place to try.**

4. To circuit-bend the card, you need to connect parts of the circuit that are not usually connected. The best results come from connecting components on the boards to a battery. To do this safely (and to avoid frying the components), you

FIGURE 4-49: **You can make a stand for your circuit by carefully peeling off the circuit board and the speaker and gluing them to a piece of cardboard. (This piece was cut out of the rest of the greeting card.)**

FIGURE 4-50: **A close-up showing mounds of hot glue protecting the wires on the circuit board. You can also see the tab cut off the original card that keeps the on-off clip from touching the metal pad below and playing the music.**

need to reduce the electric current that flows between them. A *variable resistor* lets you control the amount of electricity that can flow between two points on a circuit. The lead in a pencil (which is actually made of a soft, black substance called graphite) can conduct small amounts of electricity. That means you can use a pencil line as a variable resistor. Just take an index card and scribble a thick, heavy line along three of the edges (see Figure 4-52). Make sure there are no gaps where the pencil lines are connected. Clip one of the wires to one end of the line. Clip the other wire to the other end of the line. The farther apart the wires are along the line, the higher the resistance is between them.

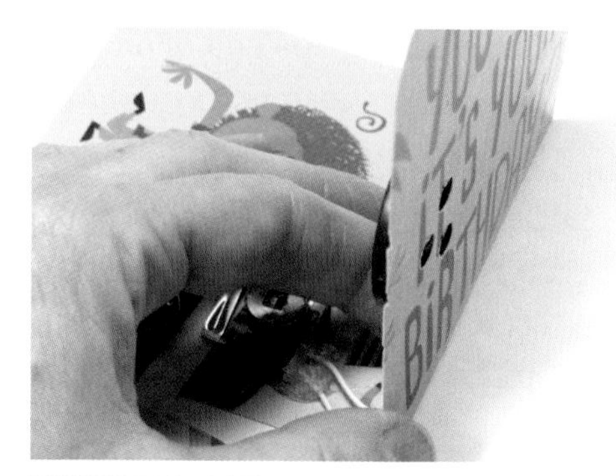

FIGURE 4-51: Use a paper hole puncher to make openings so you can hear the speaker better.

FIGURE 4-52: Make a variable resistor by drawing a thick heavy pencil line around three sides of an index card. Then attach alligator clip wires to the ends of the line.

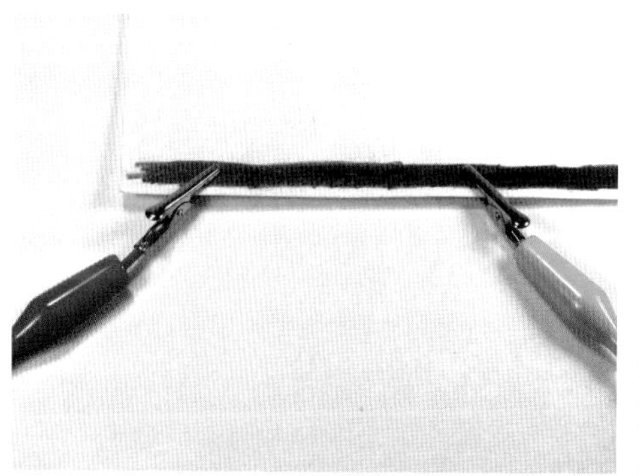

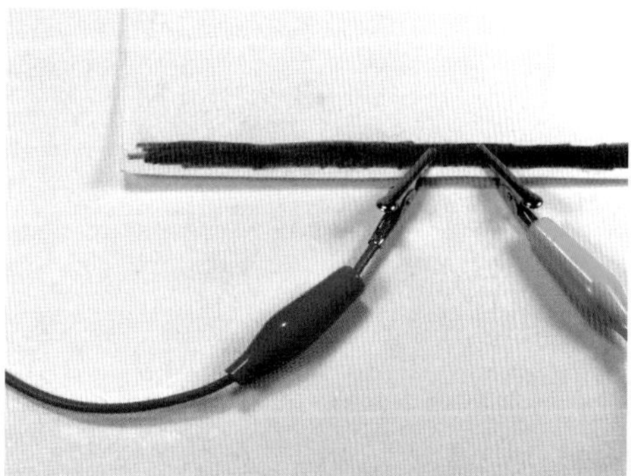

FIGURE 4-53: To lower the resistance and allow more electricity to flow through your circuit, move the two wires closer together along the pencil line.

5. Next, clip the free end of one of the wires to the metal clip that connects to a battery on the card. Use the free end of the other wire to touch different components on the circuit board. Try any interesting areas you can find. One spot to look for is a tiny rectangle marked "R," which is a resistor. You may also see other markings that show where there are other components, like capacitors, or where the speakers are attached. While you are poking around, turn on the music. If the volume or the speed of the music changes, you have found a good circuit bending point! Make a note of where it is and what happened. If you have a camera, take a close-up photo of the spot to help you remember, or shoot a video to demonstrate what happens. You can also try varying the electricity by moving one wire on the pencil variable resistor a little closer to the other wire.

FIGURE 4-54: **Attach one alligator clip wire to the on/off clip. Use the metal tip of the other to probe around the circuit board. Touch a spot on the board, then play the card to see if the sound changes. Repeat for all the metal contact points on the circuit board.**

Extension:

- Try circuit bending battery-powered toys that make noise. Spelling toys, toy electric guitars, and talking dolls are all

FIGURE 4-55: **A toy phone from the dollar store is easy to open up for circuit bending.**

favorite targets for circuit benders. Look for them at garage sales and thrift shops. Here are some ideas for what to do with them:

- Swap in a solar panel for the batteries and control the speed that way. You can remove the solar panel from an old garden light, making sure to keep the wires that are attached to it. Then remove the batteries from the toy, and connect the positive and negative terminals to the solar panel wires.
- Control the speed by using a variable resistor (like a photo sensor that varies with the amount of light shining on it) to bypass the on/off switch.

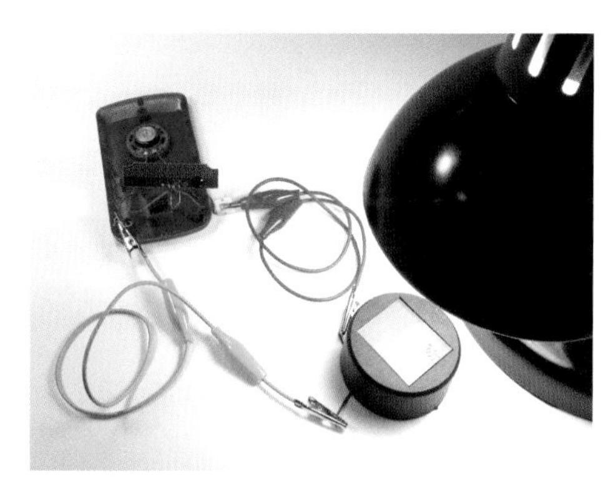

FIGURE 4-56: A dollar store talking toy phone is circuit bent by adding a solar panel from a dollar store garden light. The more light shines on the solar panel, the more voltage it produces.

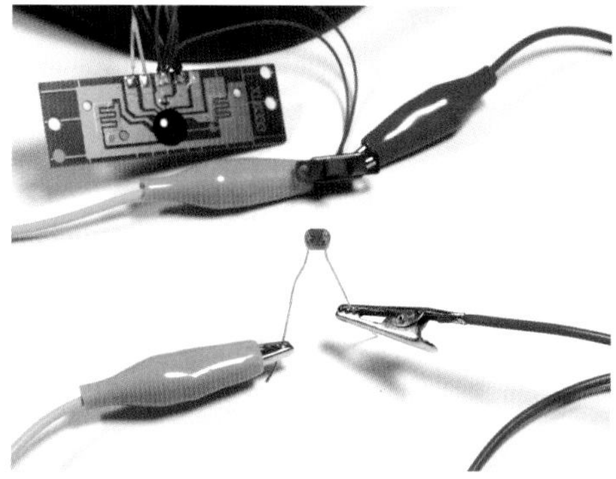

FIGURE 4-57: This light sensor is connected to the on/off switch for a toy guitar. The sensor varies the amount of voltage in the system depending on how much light it detects.

Recording and Listening 5

Recording and listening devices let you enjoy music from afar and share it with others.

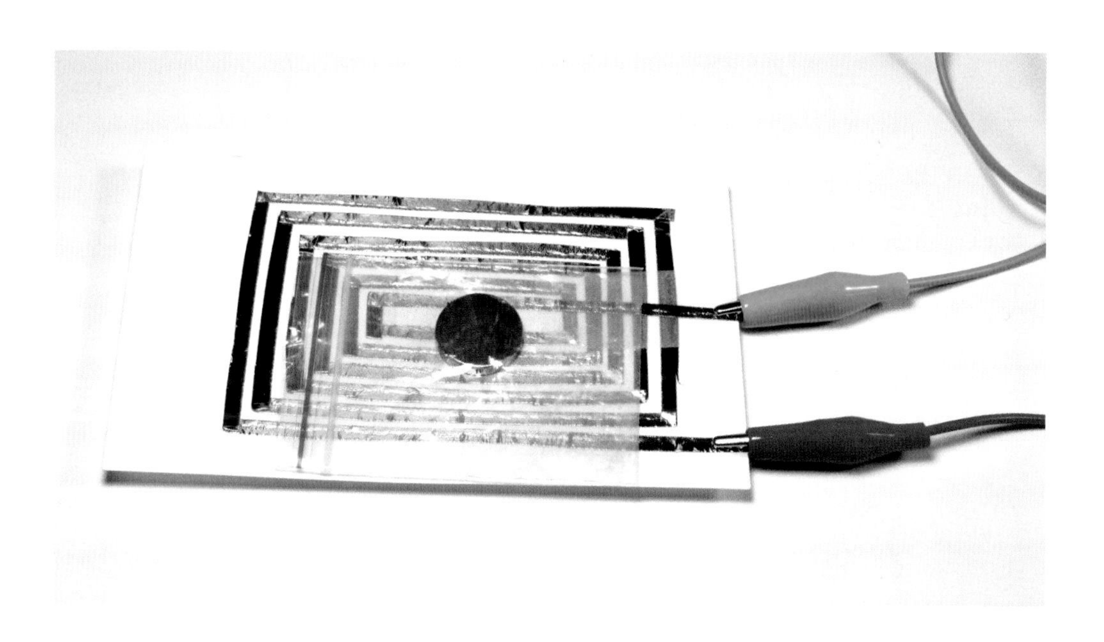

Until the late nineteenth century, pretty much the only way to hear a musical performance was to be there, live, in person. Two inventions—the phonograph and the radio—changed all that forever.

Earlier instruments let you listen to "prerecorded" performances by playing music automatically. They included wind-up music boxes (the inspiration for the project in Chapter 3) and player pianos (which used paper rolls with holes punched in them to trigger piano keys). Those kinds of machines could be programmed like primitive computers to play the same song over and over. But the phonograph and the radio made it possible to hear real musicians as they were playing. The phonograph captured a musical performance (or any other kind of sound) on a recording that you could play back later. Anyone with a record player could listen to that recording. The radio sent live performances or recorded music (as well as news, sports, comedies, and dramas) out "over the airwaves." Anyone with a receiver could tune in and hear the broadcast, anywhere within reach of the radio station's signal.

After the arrival of the phonograph and radio, instead of going out to hear a concert, you could enjoy a performance at home. Instead of having to learn to play an instrument, you could turn on a machine and create music instantly. The tradition of family sing-alongs faded. At the same time, lots of future musicians got their start by sitting alone with a record and copying their favorite performers. However, recorded music had its downside too. Listening to a song being played the exact same way, over and over, made it harder for some people to enjoy live performances, with their variations and mistakes.

The phonograph and the radio even changed the way music was made. Songs became shorter, and serious works were broken down into several parts, so they could fit on a record that was only three minutes long. Record companies and radio stations began to divide music into categories to make it easier to sell. If you were a fan of jazz or country or rock and roll, you would look for a record label or a station that featured your favorite style.

Today almost all music is stored as digital files, the same way computers store other kinds of data. Instead of collecting records, tapes, or discs, more and more people are downloading digital albums to their audio devices or listening to online subscription services. Even traditional radio broadcasts are slowly disappearing as satellites make it easier to pick up stations from around the globe. The projects in this chapter are all based on older technology that will take you back to the beginnings of music recording. You can re-create these early means of sharing music—and learn a lot about history and science in the process.

Note: The projects in this chapter require a few specialized parts and materials. Check out "The Musical Supply Closet" at the beginning of this book on page xvi for suggestions on where to find them.

What Makes Phonographs Go Round

FIGURE 5-1: A 45 rpm (revolutions per minute) record playing on a modern phonograph

Although modern record players are powered by electricity, the process of creating sound from a record is purely mechanical. In fact, the first phonographs were simple devices that you turned by hand, and they let you make records as well as play them.

A phonograph recording is basically a round object that spins around, with a long groove cut into it. As you cranked the machine, the cylinder would roll around like a rolling pin. The groove started at one end and wrapped around the

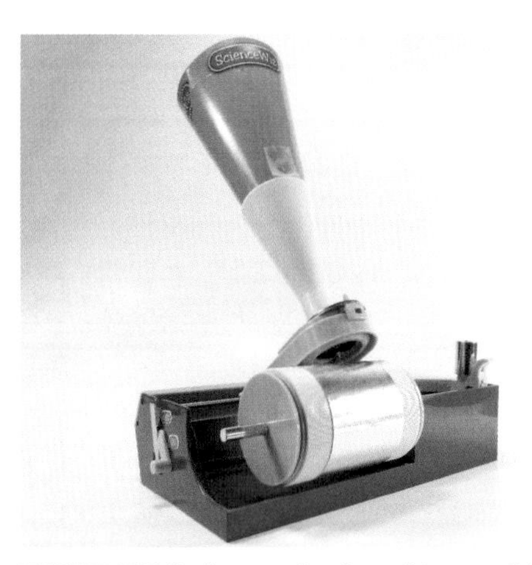

FIGURE 5-2: A motorized working model of Edison's phonograph built using the Science-Wiz Sound kit (*sciencewiz.com*). You can use it to record sounds and play them back on replaceable aluminum foil tape.

tube in a spiral until it reached the other end. A modern record is shaped like a flat disc and has grooves on both sides. The grooves on a disc start around the edge and spiral around toward a hole in the center. The hole fits onto a little knob on the record player called a *spindle*, which holds the record in place on a spinning turntable.

It's the grooves in the records—or rather, tiny bumps in those grooves—that produce the sound. To make a record the old-fashioned way, you start with a blank disk or cylinder with a smooth surface. The phonograph has a needle, also known as a *stylus*, that cuts into the record and creates the grooves. A thin membrane called a *diaphragm* connects to the needle. Sound waves directed toward the diaphragm make it vibrate. The vibrations of the diaphragm make the needle shake. And as the record spins, the movement of the needle cuts a bumpy groove in the record.

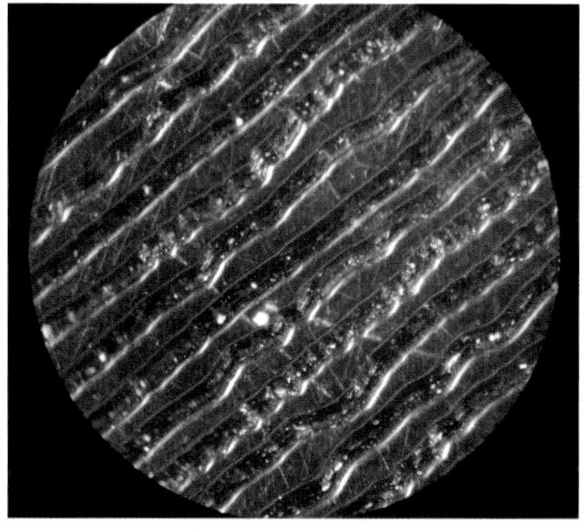

FIGURE 5-3: Grooves on a vinyl record, as seen through a microscope. The bumps are what make the needle vibrate.

Here's the amazing part: to play the record back, all you need to do is run the same process in reverse! First, you place the needle back at the beginning of the groove and start the record spinning. As the record turns, the needle rides along in the groove, and the bumps in the groove make the needle bounce along, ever so slightly. The shaking of the needle is passed along to the diaphragm and causes it to vibrate. And the diaphragm's vibrations spread to the air in the form of

FIGURE 5-4: The stylus (needle) on a modern record player is attached to a tiny strip of metal. The strip acts like a spring and lets the needle move as it hits the bumps in the grooves.

sound waves—which are exact copies of the original sounds that made the bumps in the first place!

MUSICAL INVENTORS: THOMAS EDISON

The phonograph may be simple, but coming up with the idea was a work of genius. Maybe that's why Thomas Edison called it his favorite invention. Edison was behind some of the most important inventions in history, including the electric light bulb and the movie camera. But when he created the phonograph in 1877, he thought of it as a machine for office workers to record letters to be typed up later. His early version wasn't very practical, because it used tinfoil cylinders that tore easily, so Edison set the idea aside. Meanwhile, others (including Alexander Graham Bell, inventor of the telephone) improved upon Edison's version, using wax cylinders (and later, plastic discs) instead of foil, and recorded music to play at home. Selling music records became a business in its own right, and Edison and others started their own record labels. It was the beginning of the recorded music industry, and it all started with a hand-cranked, scratchy phonograph.

Project: Manila Record Player

Materials

- Manila folder, letter-sized or larger, or a sheet of stiff cardstock or a poster board folded in half
- Old vinyl record (not a valuable collector's item)
- Pencil
- Large size belt rivets
- Very thin sewing needle (or pin or small nail)
- Two small coins
- Tape (any kind)

FIGURE 5-5: A basic record player takes just minutes to assemble.

A no-frills, nonelectric record player is easy to design. All you need is a turntable and spindle to hold the record so it can revolve smoothly, a tone arm and needle to ride over the bumps in the record's grooves, and a diaphragm to pick up the needle's vibrations and pass them along through the air in the form of sound waves. Early hand-cranked phonographs also featured a horn to direct the sound waves toward listeners.

This one-piece record player made out of a manila folder was inspired by a paper version from Instructables.com user Plugable. It's fairly loud, even without a horn to amplify it. Once you've got it working, you can invent some improvements of your own!

Warning: Playing a record on a primitive homemade record player is fun, but it *will* damage the record. Don't try this project with your family's treasured golden oldies! Garage sales and thrift shops are good places to look for old records you can pick up inexpensively and use for experimenting and inventing.

LPS VERSUS SINGLES

Vinyl records are mostly found in two sizes. The first is long-playing (LP) records, which are 12 inches (30 cm) across and have a small spindle hole. They are designed to be spun at 33⅓ revolutions per minute (rpm). Singles—also known as 45s because they are played at 45 rpm—are smaller and have a larger hole in the center. This wider hole made it easier to stack 45s on an automatic turntable and in coin-operated juke boxes. To play a 45 on a regular-sized spindle, you need an adapter. You can buy them or make your own from a scrap piece of cardboard or plastic.

1. To make your record player, start with the manila folder. One half will serve as the base of the record player, where the spindle is attached. The other half will serve as a combination tone arm and diaphragm. This half holds the needle and transfers its vibrations into sound waves in the air. If one side of the folder has ridges, use the smoother side as the turntable.
2. First, mark the spot for the spindle. Open the folder and place the LP so its edge almost touches the spine of the folder (where the folder folds). Take the pencil and trace inside the hole at the center of the record. Remove the record.
3. Take the point of the pencil and poke a hole inside the circle you just drew. Push the pencil in until the hole is as big

FIGURE 5-6: Mark the spot for the spindle by tracing the hole in the center of the record.

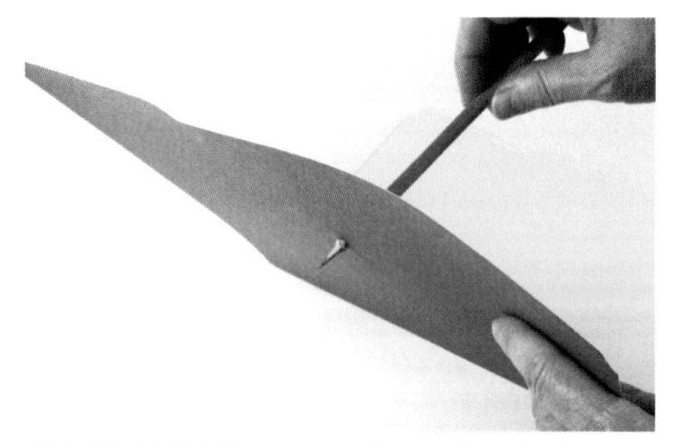

FIGURE 5-7: Use a pencil to poke a hole through the circle you drew.

FIGURE 5-8: **Use the taller half of the rivet, shown on the right.**

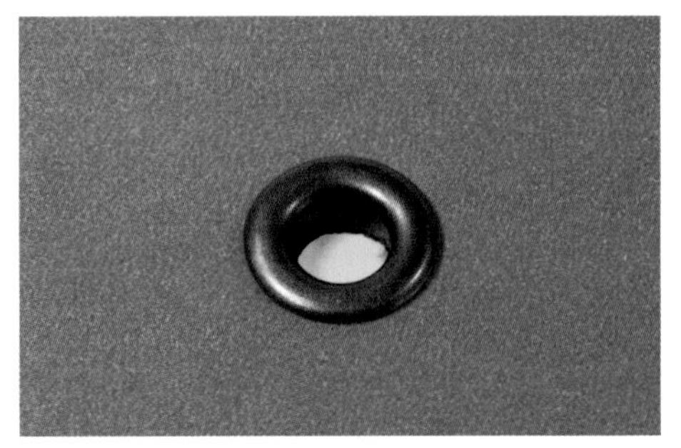

FIGURE 5-9: **The rivet as seen from the underside of the record player. It should fit snugly in the hole but still be able to turn.**

as the circle. Flip the folder over. Cut or tear away any shreds of paper from the folder around the edge of the hole. Take the rivets and find the longer, narrower half. Press it into the hole. Make sure it sticks up above the inside of the folder. This is your turntable and spindle.

4. If the front half of the folder has a tab, trim it off so the edge is straight. Then take the front and fold it in half so the edge meets the inside of the spine. Make the crease where you fold it very sharp. Draw a line to show where to make a second crease between the original crease and the edge of the folder. The two creases will bend in the same direction, about $\frac{1}{4}$ inch (1.25 cm) apart. Score the line by running something hard but not sharp along it, such as a screwdriver. Then fold along the line.

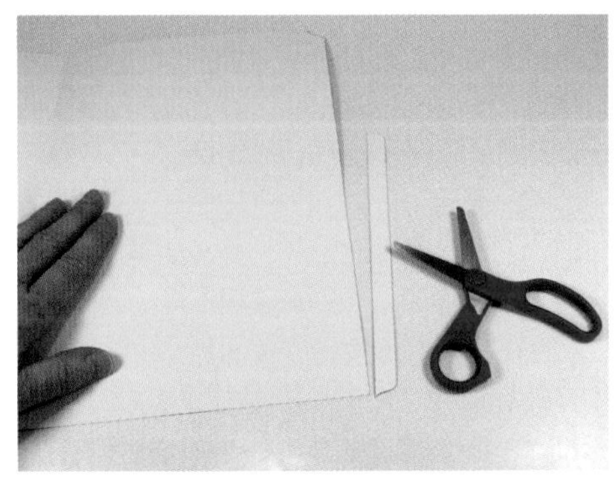

FIGURE 5-10: **If the folder has a tab, trim it off so the edge is straight.**

FIGURE 5-11: **Fold the edge down so it meets the hinge (where the front and the back of the folder are connected). Make the crease sharp.**

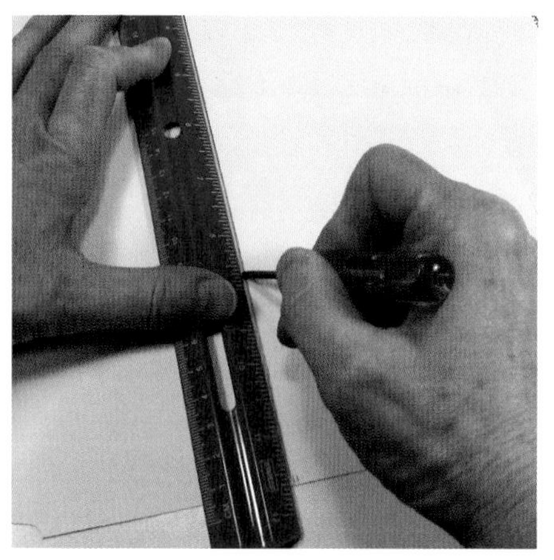

FIGURE 5-12: **Score the folder by making an indentation with a rounded point to make it easier to fold.**

FIGURE 5-13: **Make a second fold next to the first along the scored line to create a little ledge along the top.**

5. Take the needle and attach it to the center of the front half of the folder, on the outside, by pushing it in through the cardstock and out again. The point should be hanging off the edge slightly and should be tilted toward the right. Open the folder and tape two coins on either side of the needle. The added weight will help it sit correctly on the record.

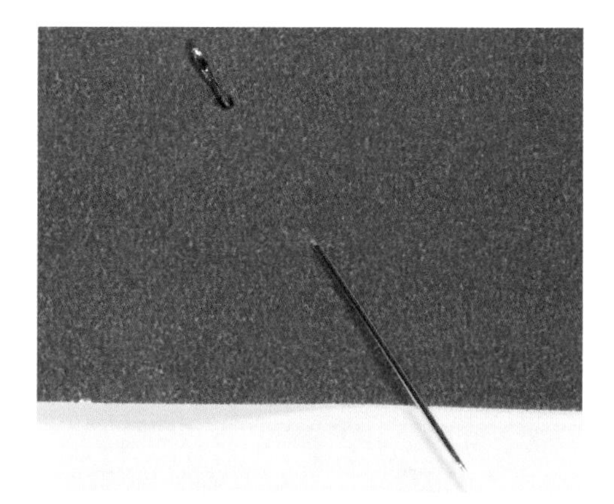

FIGURE 5-14: **Poke the needle in and out through the cardboard at an angle.**

6. To test your record player, open the folder and fit the LP onto the spindle. Adjust the needle so it rests lightly on the record. Place your finger on the smooth part of the vinyl, near the label. Spin the record around clockwise, as smoothly as possible. If the

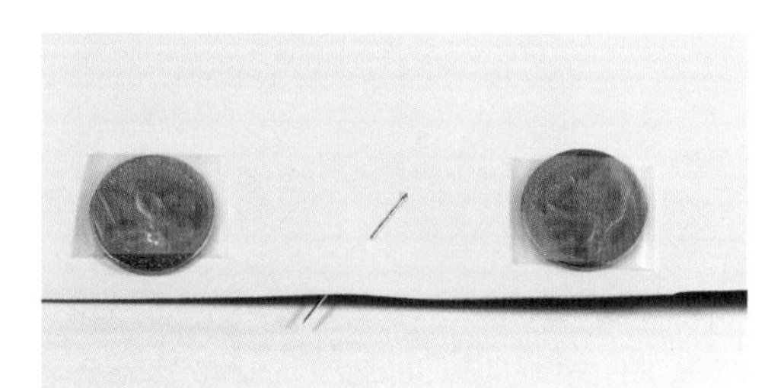

FIGURE 5-15: **Tape a couple of coins to the inside of the "tone arm" to help the needle stay in the groove.**

music—especially singing—sounds too high and squeaky, try spinning it slower. If it sounds too low and rumbly, try speeding it up.

Troubleshooting tips and extensions:

- If the folder slides around too much, rest it on a grippy surface, such as a rubber mat.
- If you have trouble turning the record with your finger, take a pencil and press it into the record using the other end as a handle.
- Try different sizes of needles, or change the angle of the needle. Tape the needle in place if you need to.
- Place your record player on top of a box, or lean a box up against the back of the cardboard tone arm, to amplify the sound with resonance.
- Turn your turntable into an instrument by using it to "scratch" a record. (See the story of Grand Wizzard Theodore that follows this project.)

FIGURE 5-16: **Lift the needle and set it gently in a groove to play a song.**

MUSICAL INVENTORS: "GRAND WIZZARD" THEODORE LIVINGSTON

In 1977, a 12-year-old from the Bronx named Theodore Livingston invented a new use for the electric phonograph. When his mom yelled at him to turn his music down, he stopped the record on his turntable from playing by holding it still with his hand. He pulled it backward and let it play, then did it over and over again. That's how Livingston—later known as Grand Wizzard Theodore—accidentally invented scratching.

Scratching, also called *turntablism*, is a way to bend old records and blend them with live vocals and other sounds. It quickly became a part of hip-hop music. And it turned DJs from disc jockeys who merely spun records at parties into musical artists in their own right.

Speaking of Speakers (and Microphones)...

FIGURE 5-17: **A mini speaker shaped like a retro boombox, a kind of portable radio/music player you carried on your shoulder**

The radio is even more amazing than the phonograph. Together with two other inventions, the microphone and the speaker, the radio turns sound waves into electricity and then back into sound waves. Today's synthesizers and electronic audio devices (like the musical greeting cards you circuit bent in Chapter 4) take the conversion process one step further, by converting electrical signals into digital information. But you can still use microphones, speakers, and radios like those invented over a hundred years ago to listen to and transmit music today.

To turn the energy of sound waves into electrical energy, you use the movement of the air to make the amount of electricity flowing through a circuit go up and down. The measure of electrical energy in a circuit is its *current*. This is how a microphone works. To change electrical energy back into sound waves, you use changes in the current to start air molecules moving. This is how a speaker works. In fact, you can use the same design as either a microphone or a speaker! A device like a microphone or a speaker that can convert one type of energy to another is called a *transducer.*

FIGURE 5-18: The speakers inside the mini boombox have copper wire coils and a plastic diaphragm. When you plug the device into a radio or music player and turn it on, grains of salt sprinkled on the speaker will jump around as the diaphragm starts vibrating.

The Super-Simple Speaker project in this chapter changes electrical energy to sound energy using a magnet. If you've ever played with magnets, you know that they have two poles, positive and negative. Put two magnets together with opposite poles facing each other and they will attract each other. Put both positive or both negative poles together, however, and the magnets will repel, or push each other away.

Now, electricity has an interesting property: whenever it flows, it also creates a magnetic field. An *electromagnet* is a temporary magnet created by running electricity through a coil of wire or other conductive material. Unlike a permanent magnet, an electromagnet can be turned on and off—it will only attract magnetic materials when the current is flowing. And you can switch the positive and negative and poles on an electromagnet by reversing the direction of the electrical current flowing through it.

MUSICAL INVENTORS: PIERRE AND JACQUES CURIE

Brothers Pierre and Jacques Curie were scientists, not instrument makers. But their discovery in 1880 of piezoelectricity led to the creation of the piezoelectric speaker. When you bang, bend, or squeeze a piezoelectric crystal or man-made material, the electrons in its molecules move around. In other words, it produces small amounts of electric current. The reverse is also true—if you zap a piezoelectric material with a small amount of electricity, it will change shape. A piezoelectric buzzer, speaker, or earphone only needs a tiny amount of current to begin moving and producing sound waves. That's why the piezo elements are used as transducers in small electronics and in crystal radios like the pizza box radio you'll be making in the next project. The Curies' discovery led to many useful inventions, including a scientific measuring instrument called a piezoelectric quartz electrometer. It was this invention that later helped Pierre Curie and his wife Marie Curie win Nobel Prize for their work on radioactivity.

FIGURE 5-19: Pierre Curie

Project: Super-Simple Speaker

Materials

Wire version:

- Magnet wire (22, 26, or 30 gauge), about 12 feet (4 m) long
- Small piece of fine sandpaper
- Wide clear packing tape

Paper version:

- Index card, 3×5 inch (7.5×12.5cm) or larger (or piece of cardstock the same size)
- Pencil
- Copper tape, ⅛ inch (3 mm) wide, about 30 inches (75 cm) long
- (Optional) clear tape

Both versions:

- Cheap pair of earbuds (that you don't mind cutting up)
- Aluminum foil
- Magnets—½ inch (1.25 cm) neodymium disks or other strong magnets—you may need to stack several together
- (Optional) alligator clip wire (test leads)
- (Optional) clear or electrical tape
- Portable device (radio, phone, tablet, mp3 player) with stereo headphone jack

Tools

- Scissors
- Wire cutters
- Wire strippers

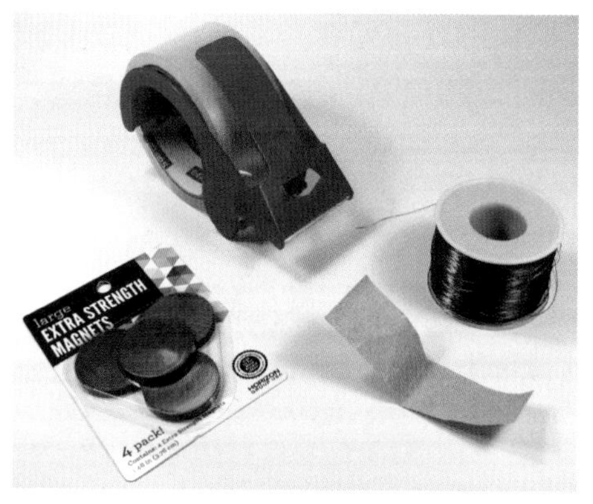

FIGURE 5-20: Materials for making a wire speaker

FIGURE 5-21: Materials for making a paper speaker

All you need to make a speaker is a conductive coil, a strong magnet, and a membrane. You don't even need batteries! For this Super-Simple Speaker project, all the electrical energy you need comes from the device you're plugged into. Speakers like this are used in musical greeting cards (like the ones you used in the Circuit Bending project in Chapter 4), earbuds, and other devices where it's not necessary to get very loud. To produce a louder volume, most speakers include an electric amplifier.

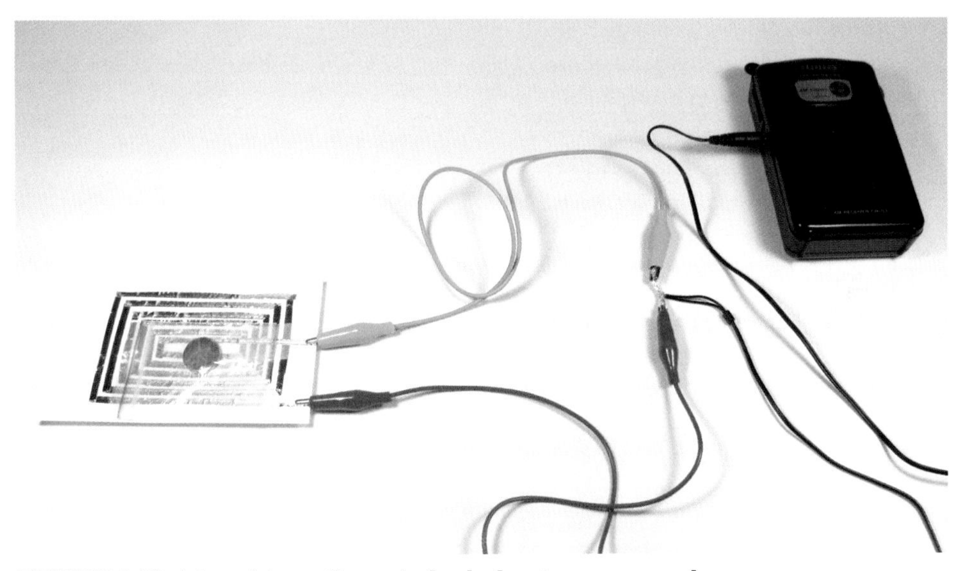

FIGURE 5-22: A transistor radio can be hooked up to a paper speaker.

The wire speakers and foil tape speakers in this project both work the same way and can be used in combination. Wire speakers are quicker and easier to make, but foil tape speakers on paper give you more opportunity to get creative. Directions for both are included in the steps that follow.

> **Note:** Speakers can be either monoaural (mono for short) or stereophonic (stereo for short) *Mono* means all the sound comes out of one speaker. *Stereo* means the music is split into speakers on the left and on the right. If you want stereo speakers, you'll need to make two coils and use two magnets (or stack of magnets). See Step 8 in the following directions for how to attach them to your earphone jack.

HOW TO STRIP WIRES

The wires you will be using in the projects in this chapter are insulated. That means they have a plastic cover or enamel coating. Because electricity can pass easily along metal wires but not the insulation, the insulation prevents the electrical charge from jumping between two places where the wire is touching and causing a short circuit. A *short circuit* is an electrical connection that occurs where you don't want it to. It can damage components or even cause a fire.

The easiest way to strip wires is with a tool called a *wire stripper*. A wire stripper looks like a pair of pliers or little garden shears, but it has one or more openings that the wire can fit into. Clamp the wire stripper around the wire (using the opening that fits best) where you want to cut the insulation—usually about 1/2 inch (1.5 cm) from the end. Then pull the wire stripper toward the end of the wire. The insulation should slip off as you pull. If you don't have wire strippers you can use a wire cutter or a pair of scissors, but be careful not to cut all the way through the wire as well as the insulation.

Magnet wire, which you need for some of the projects in this chapter, looks like shiny metal but is bright red, green, or even copper-colored. It has a very thin coating of enamel that needs to be removed. Take a piece of fine sandpaper and rub the enamel off about 1/2 inch (1.5 cm) from the end of the wire until the less-shiny copper wire inside is exposed.

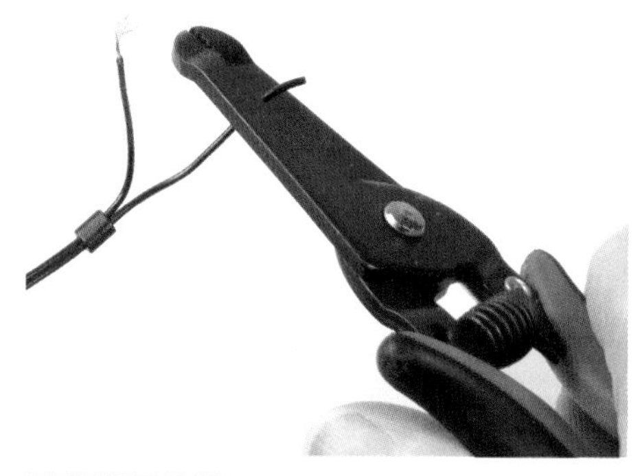

FIGURE 5-23: A wire stripper makes the job much easier.

FIGURE 5-24: Use a small piece of fine sandpaper to rub the colored enamel off the end of magnet wire.

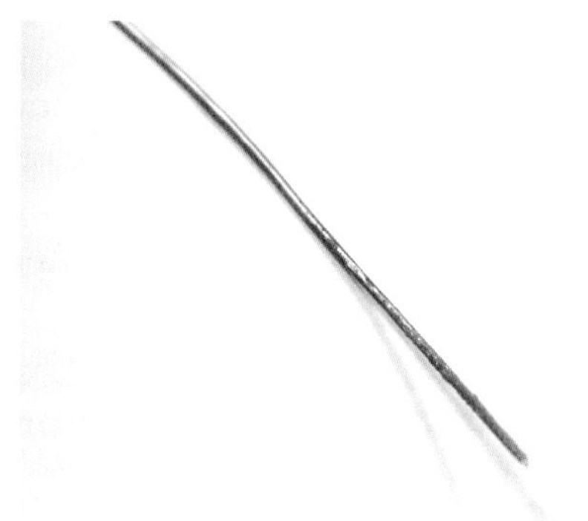

FIGURE 5-25: The darker part of the wire shows where the shiny enamel has been sanded off.

A NOTE ABOUT EARBUDS

Earbuds and some speakers use special audio jacks, a type of plug, for making connections with audio devices. If you look at the metal jack for a pair of stereo earbuds, you will see that it is divided into sections with thin black lines. The section nearest to the wire is called the *sleeve*. The point at the end is known as the *tip*. In between are one or more rings. On jacks with three sections, the sleeve is the ground, and the tip and the ring sections are connected to the right and left speakers. On jacks with four sections, the sleeve is connected to a microphone, the ring next to it is the ground, and other two sections are the right and left speakers. Make sure the jack for the earbuds you are using for the projects in this chapter matches the earphone socket on your device. If not, you may need a converter.

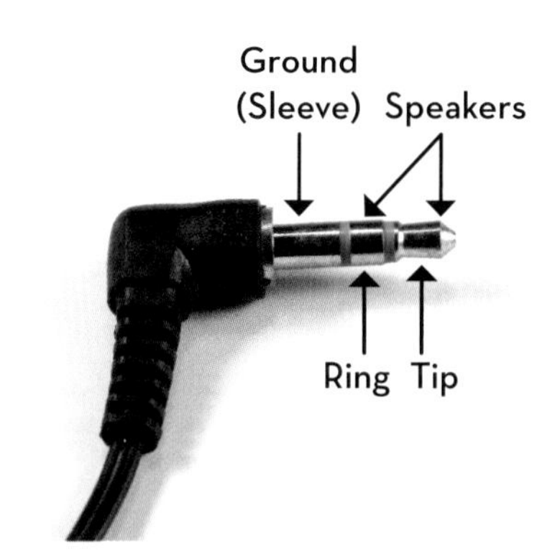

FIGURE 5-26: When clipping wires to an earphone jack, attach one to the sleeve and the other to the tip or the ring.

Safety Warning:

- To avoid losing magnets or getting them stuck together, put individual magnets in small ziptop bags, like the kind sold for beads. Be careful not to get your fingers pinched between magnets, and *NEVER* put them in your mouth or leave them where pets or small children can get to them. They can be very dangerous if swallowed.
- Use these speakers only with battery-powered radios, phones, or other music players. Do not try to use them with devices that are plugged into the wall. That could cause the wire or metal tape to heat up to dangerous levels.

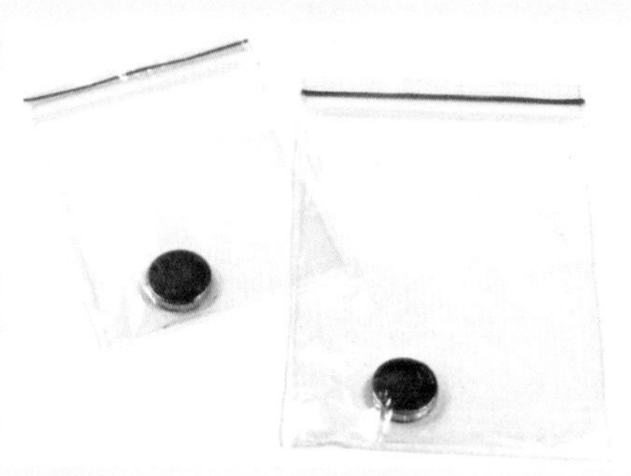

FIGURE 5-27: Put super-strong magnets in small ziptop bags for easy handling—you can use them without taking them out of the bag.

1. To make a wire speaker, start here. (Skip ahead to Step 4 to make a foil tape speaker on paper.) Take the magnet wire and coil it in a circle or oval using about 20 turns. The coils should be no wider than the width of the tape, but they can be a little longer. One easy way to make these coils is to spread your fingers apart on one hand and wrap the wire around them loosely. Leave about 8 inches (20 cm) of wire hanging loose at the beginning and at the end of the coil. It should be neat, but it doesn't have to be perfect.

FIGURE 5-28: Wrap the magnet wire loosely around your fingers to make a coil.

2. Carefully remove the coil from your fingers, keeping the loops lined up. Stick the coil onto a strip of packing tape long enough to fold over and cover the coil completely on both sides. In other words, the wires should be sandwiched inside the layers of tape. Press the two layers together, making them as tight and smooth as possible. The tape is your diaphragm.

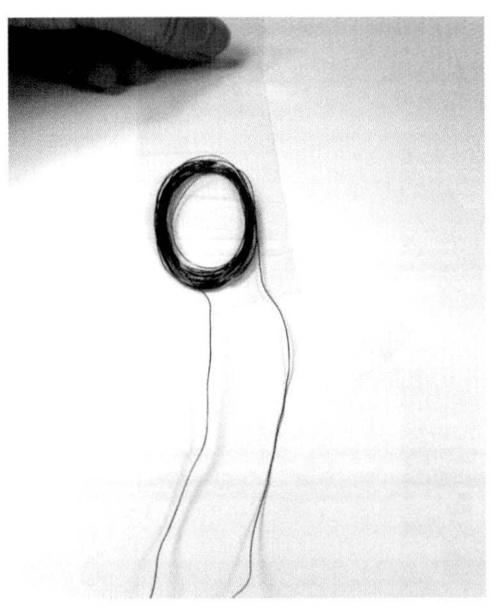

FIGURE 5-29: Attach the coil onto a strip of packing tape that is sticky side up.

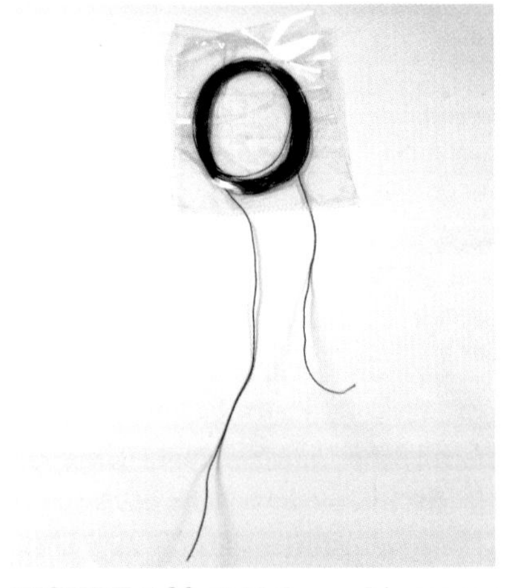

FIGURE 5-30: Fold the packing tape over the top, sealing the wire coil inside.

3. Strip the enamel off the ends of the wire by rubbing them with sandpaper. (See the sidebar "How to Strip Wires" earlier.) Then skip to Step 5 to prepare the earbuds.

4. To make a paper speaker, take an index card and lightly draw a path for your copper foil tape coil to follow. Then stick the copper foil tape right over the pencil line. (See the following sidebar, "Working with Copper Tape.") The line must begin and end at the edge of the card. (You can even wrap a little of the foil tape over the edge to the other side of the card to make a better connection.) A good design to try is a rectangular spiral. Start the circuit near one of the corners of the card and make a line going up along the edge. At the top of the coil, turn the tape without breaking it. Continue the same way, turning at each corner of the card. Keep going in tighter and tighter square spirals until you reach the center of the card. When you get to the center, you need to make a nonconductive "bridge" so you can bring the foil tape across the outside coils without causing a short circuit. To do that, place a strip of clear plastic tape over the coils you want to cross. Then stick the copper tape to the plastic tape. Bring the end of the foil tape all the way to the edge of the card. (Wrap it around the back a bit if you want.) Your paper circuit

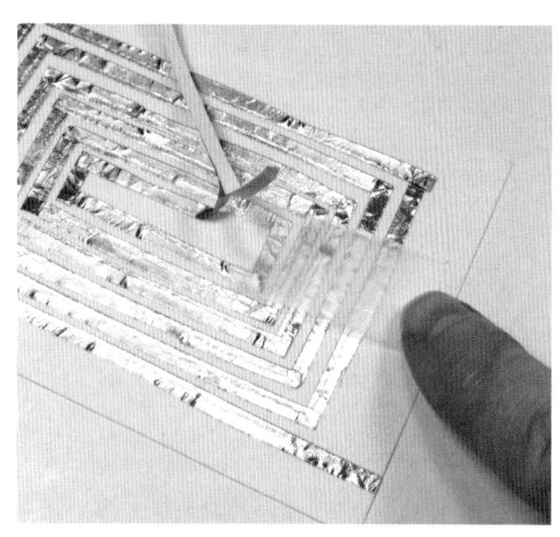

FIGURE 5-31: Make a rectangular coil with the copper tape, starting at the outside and working toward the center. When you reach the center, put a piece of clear plastic tape to make a bridge so the tape can cross over the foil without touching it.

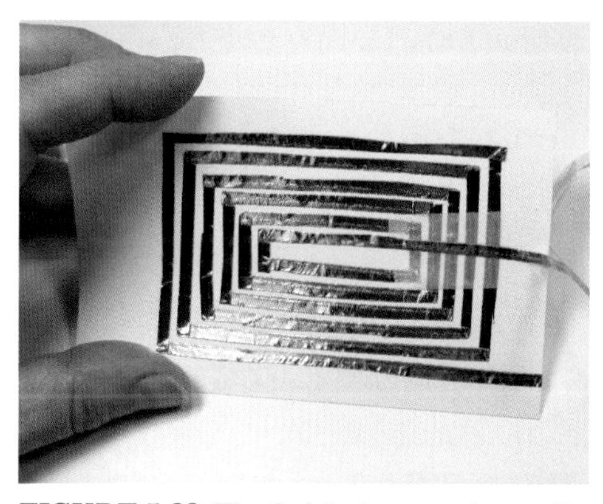

FIGURE 5-32: The finished copper tape coil

is done! Experiment with different designs—any pattern that packs the lines of copper tape closely together without short-circuiting them should work.

5. For either the paper or wire version, it's time to prepare the earbuds. Stereo earbuds usually have one wire connected to a plug that splits into two (one for each ear). Cut the two separate wires about two inches (5 cm) above the split. The remaining piece, with the earphone jack (the metal plug at the end), should be at least 2 feet (60 cm) long.

6. Strip the ends of each of these two separate wires that you cut off of the left and right earbuds. Inside the plastic housing surrounding each of the thinner wires are several wires that are even thinner. These wires must be separated. Look at them carefully. One of the wires is the ground wire. The other is the wire that brings power to that earbud. The ground wires will probably look the same for both earplugs (in Figure 3-34, both are copper colored). The left and right earbud wires will be two different colors (in the photo, one is red and one is green). Separate the thinner wires inside each earbud wire by color. Strip the ends of both the ground wire (or bundle of wires) and the colored earbud wire or wires. (If you have enamel wires that are twisted together

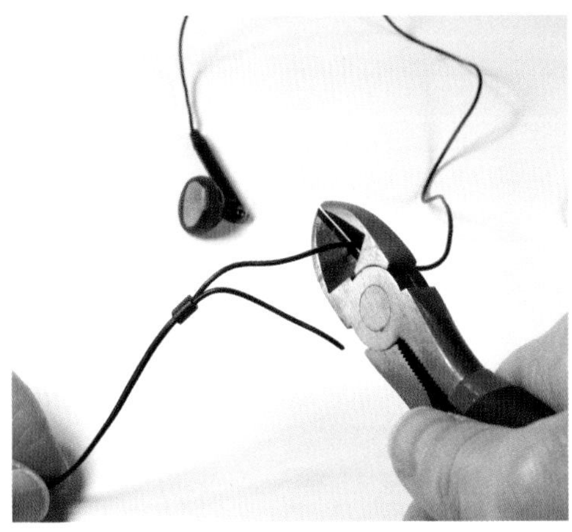

FIGURE 5-33: Cut the earbud wires, leaving a long tail connected to the plug.

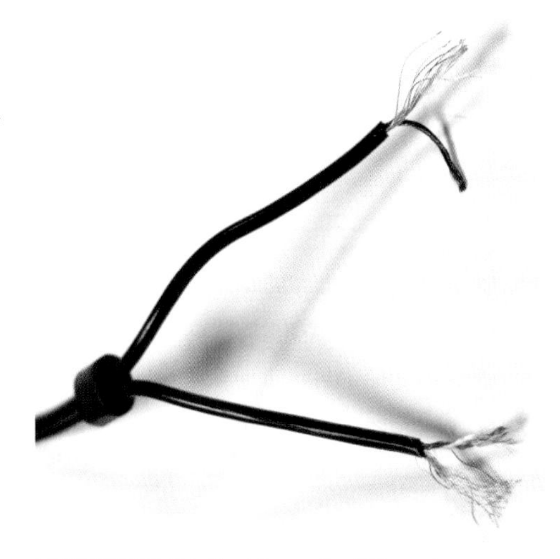

FIGURE 5-34: Inside each of the wires are even thinner wires. Separate these by color.

with string-like fibers, like the example in Figure 3-35, you must sand them very carefully. Rub the sandpaper in one direction, from the middle toward the end over and over, instead of rubbing back and forth. Keep going until the bare wire shows through the enamel coating.)

7. If you are making one mono speaker, take the ground wires from both earbud wires and twist them together. Do the same with the two colored wires. Take one of the twisted bundles of wires. Squeeze or crimp a tiny piece of aluminum foil around the wires as tightly as possible. Do the same with the other twisted bundle. Make sure the two pieces of foil do not touch.

8. If you are making two stereo speakers, wrap each of the ground wires and each of the colored wires separately with a small bit of aluminum foil, the same way you did in Step 7. You will end up with a ground and a colored earbud wire on each branch of your main earphone jack wire.

FIGURE 5-35: Strip the enamel off the inner wires very carefully by rubbing in one direction only with the sandpaper.

FIGURE 5-36: Take the ground wire from each branch of the bigger wires and twist them together. Do the same with the colored wires.

FIGURE 5-37: Wrap each bundle of wires with a little piece of aluminum foil.

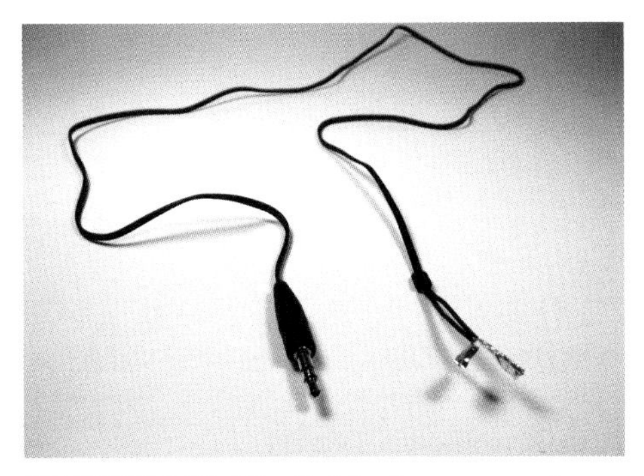

FIGURE 5-38: The finished mono earphone jack

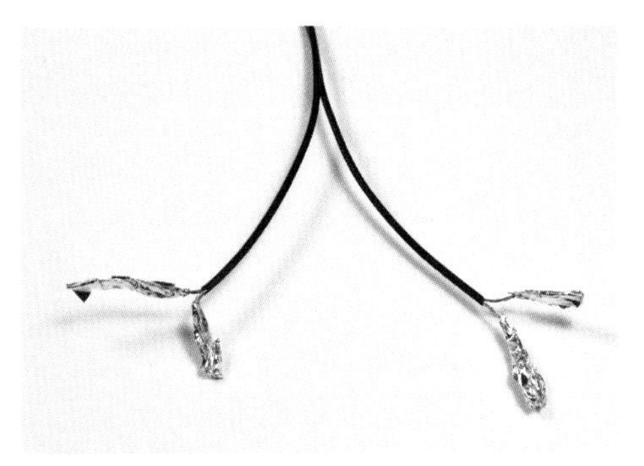

FIGURE 5-39: Wires set up for two stereo speakers

9. Use the alligator clip wires to connect the ground wire to one end of your speaker wire. Connect the colored earbud wire to the other end of your speaker wire. If you want to make a permanent connection to your speaker, twist the speaker wires around the aluminum foil covering the earbud wires and wrap them up with clear or electrical tape. Then separate the ground wires from the colored wires to form a V shape. Sandwich the whole thing in another layer of tape. This will prevent the wires from touching and causing a short circuit. For the paper speakers, you will have to make a big enough V to allow the two sets of wires to reach both ends of the foil circuit. Use clear or electrical tape to connect one set of wires to each end of the copper tape.

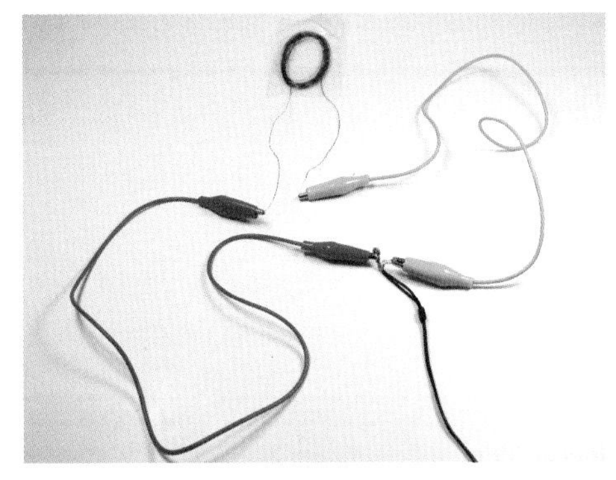

FIGURE 5-40: For a fast and easy connection, hook the wire speaker up to the earphone jack using alligator clips.

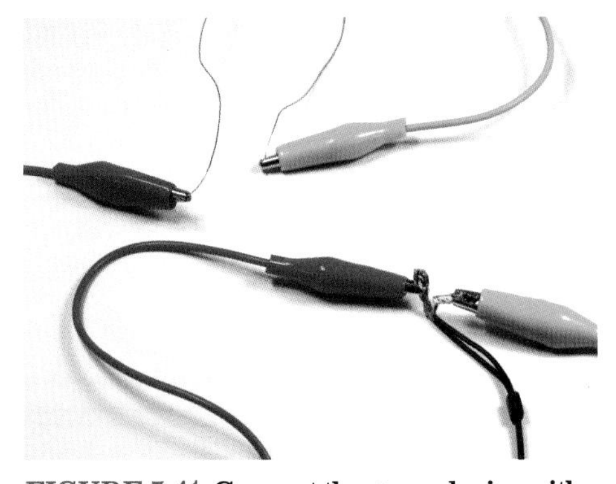

FIGURE 5-41: Connect the ground wire with one end of the wire speaker with the alligator clip wire. Do the same with the colored earbud wire and the other end of the speaker wire.

10. Time to test your speaker(s)! Place a magnet over or under the center of your speaker. Insert the earphone jack into your phone, radio, or other device and hold the speaker(s) up to your ear(s). You should hear a faint but recognizable sound coming from the speaker(s). If you don't hear anything, check all your wires to make sure there are no loose connections.

Extensions: Here are some ways to improve the sound of your speaker:

- Place it inside or on top of a resonator, such as a foam bowl or cardboard canister. If the canister has a steel bottom, you can stick the magnet right to it.
- Try different kinds of magnets, or make a stack of magnets.
- Move the magnets around to different parts of the speaker.
- Make the coil longer (more loops).

FIGURE 5-42: For a paper speaker, attach the alligator clips right to the index card, over the ends of the copper foil tape.

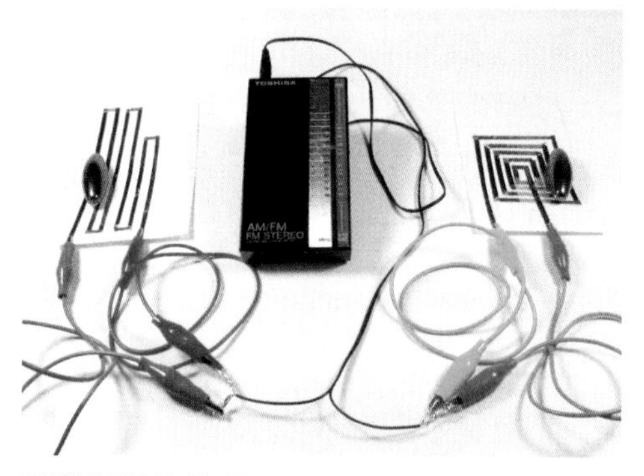

FIGURE 5-43: For stereo speakers, connect each side of the earphone jack wire to one of the speakers.

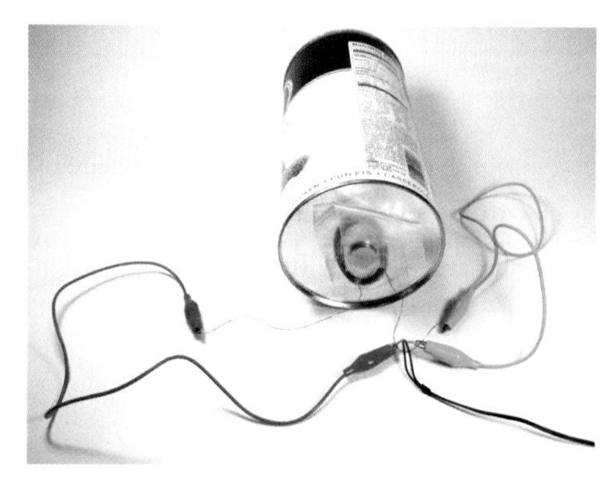

FIGURE 5-44: Add a resonator to your speaker by attaching it to a container. If it has a steel bottom, the magnet will hold your speaker on for you.

WORKING WITH COPPER TAPE

Copper tape is great for making paper circuits. Here are some tips for working with it:

- If your tape is too wide, cut it up the middle to make narrower strips.
- Leave the paper backing covering the glue until you're ready to attach it to the cardboard. Remove the backing as you press the tape down along the lines.

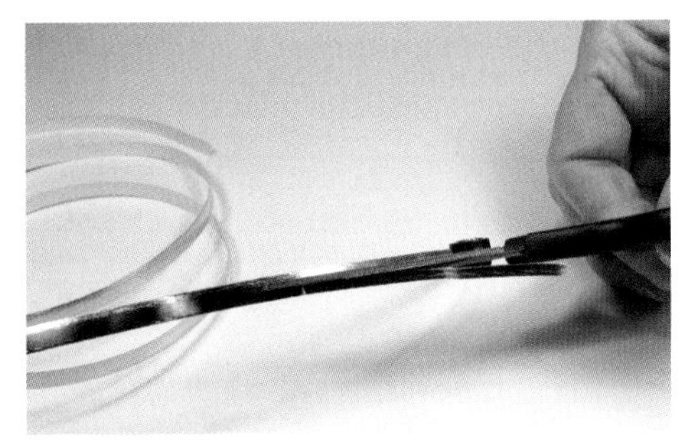

FIGURE 5-45: Cutting wide tape into narrow strips makes it easier to bend around corners and lets you fit more turns in your coil.

- To make a neat corner as you turn the tape in a different direction, fold the tape back in the direction opposite from the direction you want it to go in. Then, fold it forward, making a small pocket at the turn. Pull the paper backing off smoothly as you make the corners.
- Don't let the copper tape touch or overlap other lines of tape to avoid creating a short circuit. If you need to cross one piece with another, use plastic tape in between as insulation.

FIGURE 5-46: To bend the tape around a corner, first bend it in the opposite direction, at an angle.

FIGURE 5-47: Then fold the tape back over in the direction you want it to go.

Project: Plain Pencil Microphone

Materials

2 pencils, sharpened

Foam cup

Replacement leads for a mechanical pencil, 0.9 mm wide or thicker

3 alligator clip wires

9V battery

Earbuds

Index card

(Optional) speaker wire

Tools

Box cutter, craft knife, or pen knife

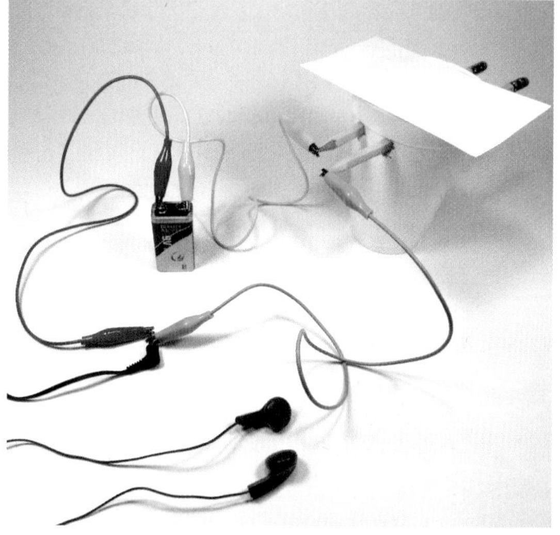

FIGURE 5-48: A microphone made from pencils needs only a 9V battery to convert your voice from sound waves to electrical signals.

Unlike the magnetic speaker, this primitive microphone uses the sound waves in the air to jiggle a loose piece of conductive material—in this case, the lead in a pencil. (Pencil "lead" is actually graphite, a soft flaky mineral that conducts small amounts of electricity.) This loose piece of pencil lead is a little like a switch that opens and closes a circuit every time it bounces around, varying the amount of electrical current that can flow through. If you send that changing electrical signal to a speaker, it translates those pulses

FIGURE 5-49: Get the thickest mechanical pencil leads you can find.

back into sound waves that copy the original sounds you made. This version of a pencil microphone was inspired by a tutorial by YouTube user Dave Hax.

Safety Warning:

- Children should get adult help to carve the wood away from the pencil. Always carve away from your hands and body.
- Hooking up a 9V battery to a homemade microphone can cause pencil leads, earphones, or other parts to heat up. Check your microphone to make sure it isn't becoming too hot, and only leave it connected for a few minutes at a time. Be sure to disconnect the battery completely before you walk away from the mic.

1. To get started, use the knife to carefully carve away a small section on one side of the pencils, in the middle. Keep going until the lead in the center is exposed. The lead should be sticking out above the wood around it.

FIGURE 5-50: Use a small knife to carefully carve the wood off the side of the pencils.

2. Take the foam cup and poke the pencils straight through, a little below the rim. The pencils should be about 1 inch (2.5 cm) apart. Turn them so the exposed lead is facing up.
3. Take a few of the mechanical pencil leads and lay them across the pencils so that they are in contact with the exposed lead.

FIGURE 5-51: Punch holes through the cups with the point of the pencils. Make sure to leave room between them.

4. Take one of the alligator clip wires and connect it to one of the pencil points. Clip the other end of the wire to the

FIGURE 5-52: Line up the mechanical pencil leads along the exposed lead in the wooden pencils.

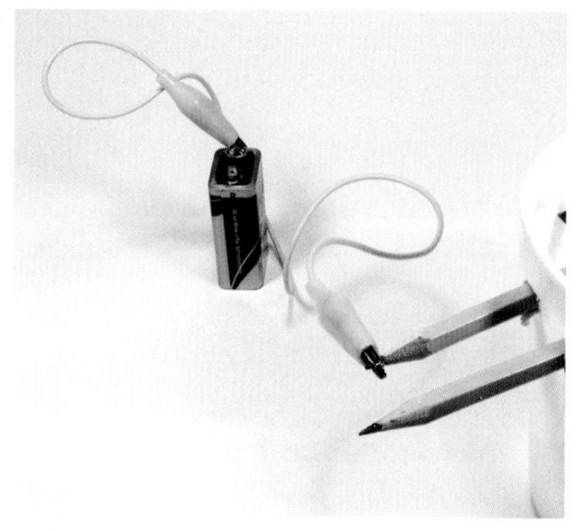

FIGURE 5-53: Attach the first alligator clip wires to one of the pencil points and the negative terminal on the battery.

negative terminal on the 9V battery. (You should see "+" and "–" markings on the battery to indicate the positive and negative terminals.)

5. Take another alligator clip wire and attach it to the positive terminal of the battery. Connect the other end of the wire to the tip of the earphone jack (see "A Note about Earbuds" earlier in this chapter).

6. Take the last alligator clip wire and connect it to the ground segment of the earbud jack (the part nearest to the wire). Clip the other end to the point of the other pencil.

7. At this point, your microphone should work. Try talking or singing into it to see if the sound can be heard through the earbuds. Then rest the index card

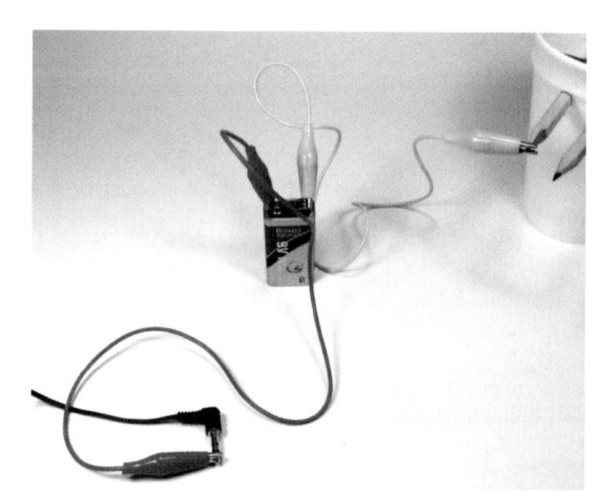

FIGURE 5-54: The second alligator clip wire connects the positive terminal of the battery and the tip of the earphone jack.

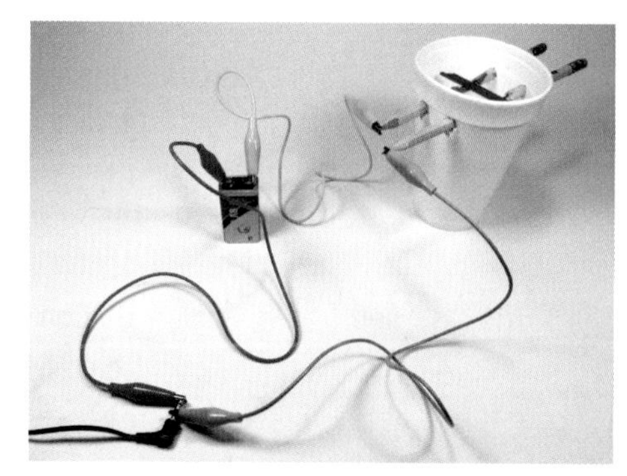

FIGURE 5-55: With the third alligator clip wire, connect the sleeve of the jack to the other pencil point.

on top of the cup to serve as a diaphragm. See if that makes the microphone sound better.

Extensions:

- Your microphone will pick up some kinds of sounds better than others. High or low frequencies may be harder to hear, and some letter sounds may not be audible. Experiment to see what comes through and what doesn't.

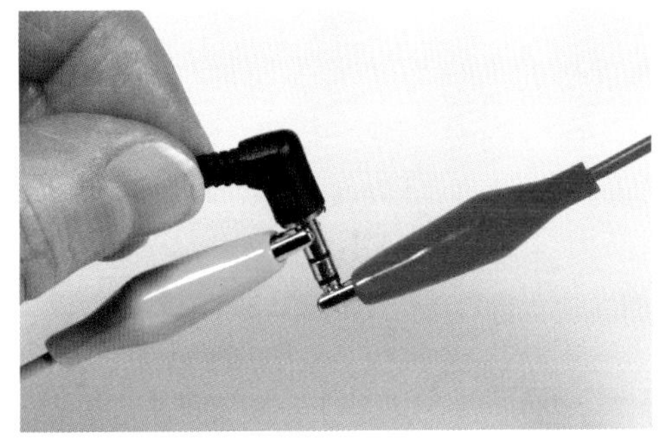

FIGURE 5-56: A close-up of the earphone jack with the wires attached

- Try different materials to make the connection between the pencils, such as a paper clip or a thin nail. Make notes of which materials work best.
- Replace the index card with different kinds of diaphragms to see if you can improve the quality or range of the microphone. For example, try stretching a balloon or piece of plastic wrap across the top of the cup.

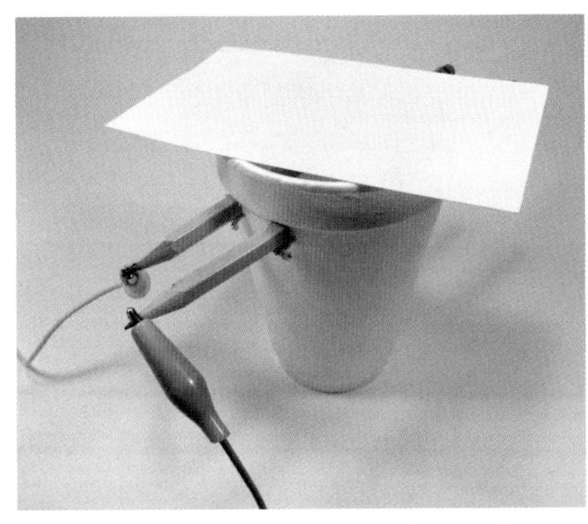

FIGURE 5-57: Test whether adding a diaphragm such as an index card improves the sounds picked up by your microphone.

- Use speaker wire to extend the distance between the earbuds and the microphone. Clip the alligator clip wires to each of the speaker wire ends, and use additional alligator clip wires at the other end to attach the speaker wire to the earphone jack. Get a helper to go into another room with the earbuds and see if they can still hear you transmitting into the microphone.

Tuning into Radios

Radio waves are a form of electromagnetic energy. They can travel through the air, like sound waves, but they can also travel through the vacuum of space. In fact, some scientists use giant radio telescopes to collect radio waves given off by distant stars.

FIGURE 5-58: Antique crystal radio enthusiast Chuck Porter picks up a signal.

A radio has an antenna—a metal rod or wire—that picks up the electromagnetic energy and turns it into electrical current. But the antenna picks up radio waves from all the stations in your vicinity (and on good nights, even stations from distant cities). In order to listen to just one station at a time, you need a tuning circuit. Part of this circuit consists of a wire coil. The electricity flowing through the coil creates a magnetic field, just like in a speaker or microphone. A coil or component that stores energy in the form of a magnetic field is called an *inductor*. Along with the coil, you also need a capacitor. A *capacitor* stores energy in the form of an electrical field. A *variable capacitor* lets you adjust the amount of charge it can store. Together, the inductor and the variable capacitor make the tuning circuit resonate at one particular frequency, which amplifies radio waves at that frequency, letting you zero in on the station you want to hear. (A capacitor and an induction coil can also be combined to serve as an oscillator. That's the device that generates electrical signals with different frequencies in a music synthesizer.)

Once you've singled out the frequency you want to hear, you need something to turn the electrical current back into sound waves. The first step is to get them flowing in the right direction. Radio waves make electrons in the circuit flow back and forth, and the high points and low points cancel each other out—a little like the way reflected sound waves merge to create a standing wave. So you need to keep the current from flowing both ways. A *diode* is an electronic component that only allows electricity to flow in one direction. If you look at the waveform of an electrical current going through a diode, it will appear as if the bottom of the wave has been cut off, leaving only the part from the midpoint up. In early radios, the signal was filtered by sending it through a thin piece of wire, referred to as a *cat's whisker*. The cat's whisker rested lightly on a chunk of mineral such as galena, a type of shiny gray crystal, or even a rusty piece of metal. Radios that run off the energy of the radio waves themselves, and use diodes, raw mineral, or strips of metal to filter their signal, are known as *crystal radios*.

You now have pulses of voltage that can be used to make an earphone or speaker vibrate and create sound waves in the air.

Since crystal radios use only the energy of the radio waves themselves, you need very sensitive earphones, like piezo earphones, or speakers with electric amplifiers, such as battery-powered speakers, to hear anything. To complete the circuit (remember, electricity only flows when it has a closed

FIGURE 5-59: A rock containing galena, the mineral used in the cat's whisker diode (directly above the rock sample) on an antique crystal radio set

circuit to travel along) most crystal radios also need a ground wire. However, the following crystal radio project can work even without a ground wire, making it portable.

MUSICAL INVENTORS: GUGLIELMO MARCONI AND NIKOLA TESLA

Italian physicist Guglielmo Marconi is usually given credit for the invention of the radio. But another famous scientist—Nikola Tesla—actually patented the idea first. Tesla developed the induction coil, which is needed to send and receive radio waves, in 1884. A year later, Marconi transmitted a radio signal in England. Both men applied for a U.S. patent, which would give them ownership of the invention. In 1900, the government sided with Tesla. But Marconi had some rich and powerful backers (including Tesla's rival, Thomas Edison) and founded his own radio company anyway. In 1909, Marconi was awarded the Nobel Prize for the invention of the radio. Tesla was furious. The U.S. Supreme Court upheld Tesla's claim in 1943, but it was too late—Tesla was dead, and Marconi's name went into the history books as the inventor of the radio.

FIGURE 5-60: Guglielmo Marconi

FIGURE 5-61: Nikola Tesla

Project: Pizza Box Radio

Materials

- Pizza box, preferably unused (the largest you can find) or a similarly shaped cardboard box
- 4 small plastic or paper cups (use sturdy ketchup cups from a take-out place, or cut down a small cup to about 1 inch [2.5 cm] high)
- (Optional) white glue (if using paper cups)
- Small piece of fine sandpaper
- Tape (either clear, masking, or electrical)
- 4 brass brads
- Paper plate (large)
- Wax paper
- Aluminum foil
- (Optional) aluminum foil tape
- (Optional) copy paper
- 1 germanium diode (1N34A)
- 1 crystal radio (piezo) earphone
- 2 pieces of insulated wire, any kind, about 15 inches (38 cm) long (preferably different colors—or you can just use more magnet wire)
- (Optional) alligator clip test lead wires
- (Optional) extra antenna wire and/or ground wire 10–50 feet (3–15 m) or more magnet wire or any kind of wire

Tools

- Iron
- (Optional) hot glue gun (for plastic cups)
- (Optional) wire cutters
- Wire stripper

This home-baked AM radio is a variation on the standard crystal radio. All the components fit in a pizza box, which means it's portable enough to carry around with you! The secret: the inductor coil also acts as the antenna. This kind of loop antenna radio also works without a ground wire, because the loop itself makes a complete circuit. But because it doesn't have

FIGURE 5-62: Ask your local pizza place for a clean box—as large as possible.

an outside antenna, this radio works best within a couple of miles of a radio transmitter. For better reception, you can easily hook it up to a ground wire and an antenna.

Safety Warning: If you add an antenna or ground wire to your radio, avoid touching other electrical wires. If you are outside, remove the wires when you are done. Wires strung in trees or on the outside of a building can cause damage during a thunderstorm by conducting lightning back to you!

Note: See "The Musical Inventions Supply Closet" at the beginning of the book for tips on where to find specialized parts.

FIGURE 5-63: If you are close enough to an AM radio transmitter, the pizza box radio may pick up a signal without an antenna or ground wire.

1. On the pizza box radio, the wire coil is wound around the inside of the box. The rest of the parts are attached to the outside of the box, near the edge where the lid is hinged. Since you will hold the pizza box with the hinge side up, you can call this side the "top" of the radio. The lid and the bottom of the box will become the front and the back of the radio. To get started, put the box down the normal way and open the lid with the hinge away from you. Make sure the cups you are using are shorter than the walls of the box. If they are not, cut them down to size. Use white glue (for paper cups) or hot glue (for plastic cups) to attach the cups to the bottom of the box, near the corners. Leave a little space between the cups and the corners so you have room to wind the wire around them. Make sure the cups are attached securely.

2. Before you begin to wind the wire, leave about 1 foot (30cm) of wire hanging loose. Start at the right-hand corner nearest the top of the radio (the lid hinge), and bring the wire down along the right side. Bend the wire around the cup at the right corner nearest you (the bottom of the radio). Continue winding the wire around the cups the same way. Try to keep your coil neat by stacking each level of wire on top of the one before it, but it's OK if the wires overlap. You can use a few pieces of clear tape to hold the wire in place. Wind about 60 feet (20 m) of wire around the box, ending at the upper left-hand corner. Leave about 1 foot (30 cm) extra at the end of the wire before you cut it.

FIGURE 5-64: Attach the cups to the box securely with glue, then wind the copper wire around them neatly to make the antenna/coil. Hold the wire in place with tape.

Note: Before you start winding the wire, measure the side of your box to get an idea of how many loops you will need. For a medium-sized pizza box that can fit loops that are 1 foot (30 cm) on a side, you will need to make 15 loops (four sides times 1 foot per side equals 4 feet per loop, and 60 divided by 4 is 15). If your box is larger, you will make fewer loops. If you need to, you can add wire by twisting the ends of the old piece and the new piece together. Before you do, however, don't forget to strip off any insulation (or sand off the enamel insulation if you are using magnet wire). You need to make sure that the bare metal ends of both pieces of wire are touching as you twist them. Secure them with electrical tape.

3. When you are done with your coil, you will have an extra piece hanging off each end. Feed each end through the opening in the hinge to the lid nearest to its corner (or make your own openings if there aren't any). Tape the wire down as it leads to the hole. Then poke three small holes through the top of the radio, near the center. The holes should be about 1½ inches (3.75 cm) apart. From the outside of the box, push a brad through one of the holes. Leave a little space between the head of the brad and the cardboard so there will be room to attach wires or clip a test lead. On the inside of the box, bend the legs of the brad to hold it in position. Secure the legs to the cardboard with tape. Repeat with the other two brads. Close the lid. Make sure the lid doesn't disturb the coil wires.

FIGURE 5-65: Insert three brads into the top of the radio from the outside. Raise them a little above the surface before bending the legs out and secure them with tape. (Note: If you follow the directions, your wire ends will not cross each other as they do in this test version and will look much neater!)

FIGURE 5-66: Fasten the brads to the box so that the heads stick up a little above the outside of the box. This gives you room to attach wires to the necks of the brads.

4. Next, make the capacitor. It consists of two smooth, flat pieces of aluminum foil, with wax paper separating them so they don't touch. One piece slides over the other. The more they overlap, the higher their capacitance. To start making the capacitor, place a sheet of aluminum foil between two sheets of wax paper. With an iron set on low (no steam), iron the wax paper and foil flat. Take the sandwiched sheets and lay a large paper plate on top of them. Trace around the plate, and add a tab sticking out of the side that is about 2 inches (5 cm) square. Cut out the shape. Then cut both circles in half, so that the tab is also divided in half. *Important:* Be sure to leave some of the foil on the tab exposed.
5. Now lay one half of the capacitor on the center of the lid. Tape the shape to the box, with the tab sticking out toward the side of the box. Try to keep it as smooth and flat as possible. This will be the lower half of the capacitor. To make the upper half, cut the paper plate you used to trace the capacitor shape in half. With white glue or a glue stick, attach the

FIGURE 5-67: **Sandwich a piece of aluminum foil between two sheets of wax paper and iron it on low heat to make them stick together.**

FIGURE 5-68: **Take the wax paper/foil sandwich and trace a large circle, using the paper plate as a guide.**

FIGURE 5-69: **Before you cut out the circle, draw a long tab hanging off one side. Cut the circle in half, right down the middle of the tab.**

other half of the foil/wax paper sandwich to the bottom of the plate, matching the edges as much as possible. Tape them together around the edges for extra strength. (*Optional:* To make the tab on the upper half of the capacitor a little sturdier, back it with some extra cardboard cut from the remaining half of the plate.) With the plate side up, lay the upper half of the capacitor on top of the lower half. Line the two half circles up so the straight edges are matching. Find the middle of this edge and go a little bit down, toward the rounded side. Poke a hole through both halves of the capacitor with a pencil point or other pointy object. The hole should go all the way through the lid of the box as well. Make sure the hole is big enough to allow the upper half of the capacitor to turn easily when you fasten it on with a brad. Then insert a brad through the hole and fasten it on the inside by bending the legs open and taping them to the inside of the lid.

6. Take the two pieces of insulated wire and strip off the insulation at the ends. (Don't forget that if you are using

FIGURE 5-70: Tape the lower half of the capacitor to the lid of the pizza box. Cover all the edges with tape, but leave some of the foil on the tab exposed.

FIGURE 5-71: Attach the upper half of the capacitor to the lower half with a brad. Rotating the upper half like a dial will let you tune the radio to different frequencies.

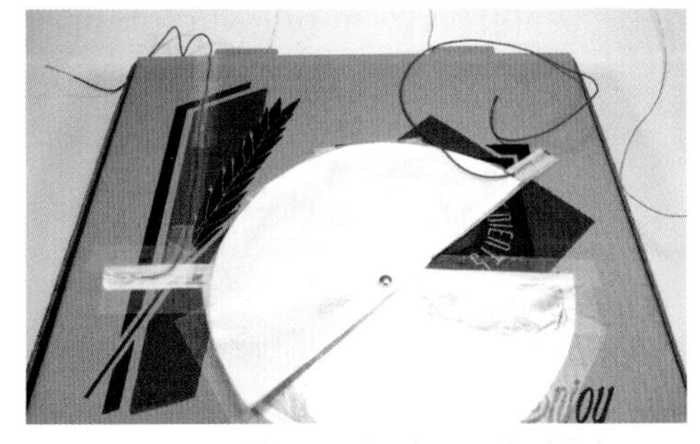

FIGURE 5-72: The black wire on the left is taped to the tab on the lower half of the capacitor. It is attached to the lid with more tape. The red wire on the right is taped to the tab on the upper half of the capacitor, but the rest is left loose.

magnet wire, you need to sand off the enamel to expose the bare copper underneath.) Tape one end to the tab on the lower half of the capacitor so that the bare wire is touching the exposed foil. Bend the wire toward the top of the radio and cover it with tape to hold it in place. Attach the other piece of wire to the tab on the upper half of the capacitor in the same way—*but do not tape it to the box!*

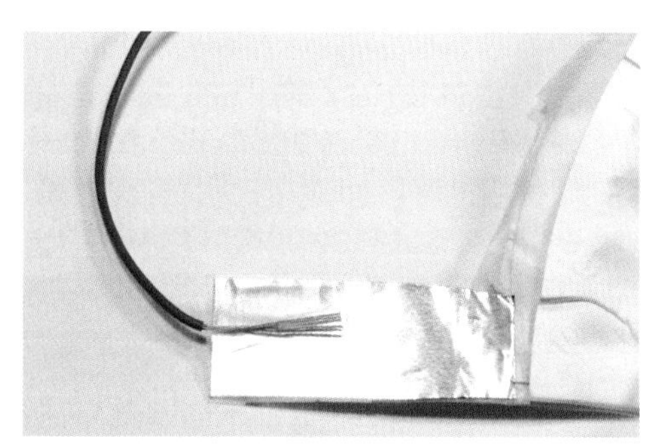

FIGURE 5-73: Make sure the metal end of the wire is touching the metal foil on the tab.

7. Grab one of the free ends of the magnet wire from the coil/antenna sticking out of the holes at the top of the radio. Strip off the enamel insulation by sanding it. Take the capacitor wire that is closest. Twist their bare metal ends together. Then wrap the twisted ends around the nearest brad (not the middle brad). Repeat with the other magnet wire from the coil and the other capacitor wire.

FIGURE 5-74: Sand off the enamel insulation on the magnet wire, and then twist the end together with the end of the nearest capacitor wire.

8. Take the germanium diode and bend the wire near the end with the black stripe around the brad in the middle. Bend the other end around one of the outside brads. Check to see that the diode wire is touching the brad and/or the wires on both ends to make a good electrical connection.
9. The final step is to connect the earphone. Although there is only one ear piece, there are two wires. The electric current

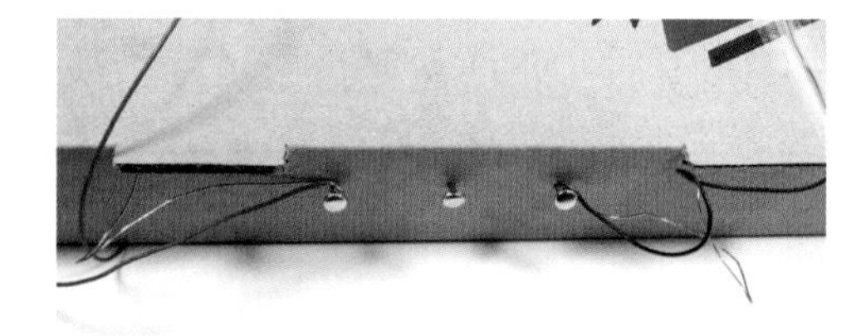

FIGURE 5-75: Wrap the twisted ends of the wire around the nearest brad.

flows into one wire, through the earphone, and out the other wire. (It doesn't matter which.) If there isn't a lot of bare metal wire sticking out from the earphone wires, you may need to strip the ends. Be careful because they may be made up of multiple strands that are very thin and breakable. If they are, you can twist the multiple strands together to make them a little easier to work with. Wrap one of the earphone wires around the middle brad. Wrap the other wire around the outside brad opposite the diode.

10. Your Pizza Box Radio is ready for testing! A loop antenna radio works best when the loop is facing the direction of the radio waves. If you know of a nearby radio transmitter, start by facing in that direction. Place the earphone in your ear and hold the pizza box with the top—the side that has the brads and all the wires connected to it—facing up. It may help to hold the back of the radio against your chest. Place one hand on the capacitor. Slowly turn the upper piece of the capacitor like a dial and try to tune in a station. Chances are, if you find one, it will be very faint. If you do find one, mark it with a pencil. If you want to look for more, or if you don't get a clear signal on the first station, turn the capacitor a little and try fiddling with the capacitor dial again.

FIGURE 5-76: The germanium diode and earphone complete the circuit.

> **Note:** You can search for AM radio stations in your area on the website *radio-locator.com*. It tells you the distance from your location to the radio station, the strength of the signal, and the format (kind of music or content). If you click on the information button for an individual station, it also tells you the exact location of that station's transmitter (the tower that broadcasts the signal) with a link to its position on Google Maps. That can help you figure out which direction to face in when you are trying to get a signal on your radio.

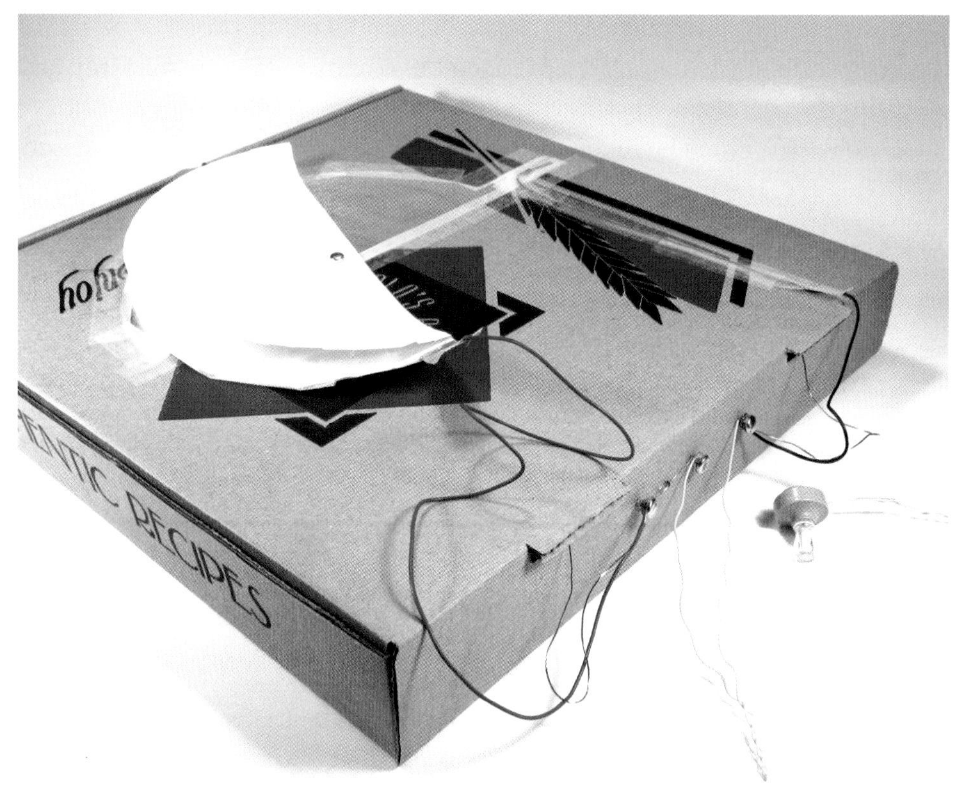

FIGURE 5-77: **Your radio is complete and ready for testing.**

PIZZA BOX RADIO TROUBLESHOOTING TIPS

Getting any crystal radio to work, including the Pizza Box Radio, can take a little work. Weather conditions (on Earth and in space), nearby buildings, and the shape of the landscape are just some of the factors that can affect your ability to receive signals. And if you remember the DIY Theremin project in Chapter 4, you know that every time you come near the radio, the capacitance of your own body changes the tuning a little bit!

If all you hear is static—a scratchy or buzzing noise—that's still a good sign. It means your radio is picking up signals, even if they're not radio broadcasts! It may even be an appliance that's suddenly drawing a lot of electricity that's causing the crackle in your earphones, but it's proof that your circuit is working.

Here are some tricks that may help you pick up a radio signal:

- If you're not hearing anything at all, not even static, try touching or tapping the capacitor or wires to see if doing so makes your reception better or worse.
- Disconnect one of the headphone wires and tap the metal end on the brad or wires. You may get a burst of static that will crank up the radio's receiving power.
- If you're inside, move near a window, or just try different parts of the building. You may also have better luck outside than inside.
- AM radio signals travel better at night. (It has something to do with changes in the atmosphere.) Try listening after dark to see if you can pick up more stations and stations that are farther away.
- The number of loops your coil has affects how large a range of frequencies your box will pick up. If you aren't getting any signals, you can try making your wire longer or shorter.

The best way to improve your radio's reception is to hook it up to a ground wire and/or an antenna. If your building has an antenna wire coming into the room, take an alligator clip wire and connect one end to the antenna and the other to one of your coil wires. If you don't have access to an antenna wire, you can try connecting to any large metal object that is not touching the ground (or one that is insulated from the ground, like a bicycle with rubber tires).

You can also make your own antenna with a piece of wire. The longer the wire, the more energy it will pick up, and the louder the station will sound. Inside, you may want to string the wire down a long hallway or go from room to room. If you are outside, you can drape an antenna wire on a tree (but always remove it when you are done, to avoid attracting lightning!).

A ground wire can be connected to anything metal that goes into the ground, like a metal fence post. Inside, you can tape a ground wire to a cold water pipe or faucet. Make sure that the pipe is all metal between the point where you connect your ground wire to the place where the water pipes go under the earth. (Some water pipes are PVC plastic. Hot water pipes won't work because they are connected to the hot water heater, not the underground water main.)

FIGURE 5-78: One way to ground your radio's circuit is to attach a wire to a cold water pipe under a sink.

For easier listening, you can also replace the earphones with a battery-powered speaker that has an earphone jack. Use alligator clip wires to connect the ground portion of the jack (the sleeve, nearest the wire) to the end brad. Use a second alligator clip wire to connect the middle brad to the tip of the earphone jack.

Connecting your Pizza Box Radio to an antenna and ground wire will limit the effect of pointing the coil toward a station. And of course, it won't be as portable. But it will make any signals you pick up louder—sometimes loud enough to hear quite easily. So if you're having trouble, give it a try!

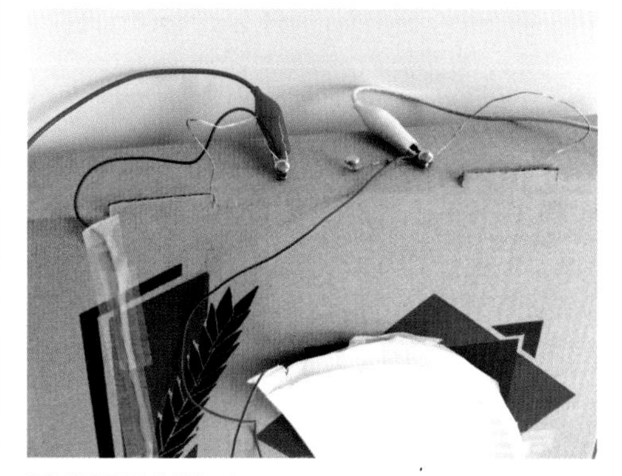

FIGURE 5-79: An antenna wire and a ground can be clipped on or twisted around the outer brads for better reception.

FIGURE 5-80: You can use an amplified speaker instead of an earphone by connecting the sleeve and tip of the speaker jack to the same brads you use for the earphones.

FIGURE 5-81: Listening to a pizza box radio outside using a speaker (hidden in the shadow between the two listeners)

Afterword: It's Not Music Until You Make It Musical

As you learned at the beginning of this book, music is a series of sounds that has pattern, repetition, and meaning. The projects in this book show you how to create all kinds of interesting sounds: tweets, blatts, pops, slides, bangs, buzzes, whistles, and pure ringing tones. But to make them musical, you have to shape and arrange them, to give them a form that listeners can recognize and understand.

This is what musical inventors have done since the first cave dweller blew into a hollow bone. And it is what musical inventors do today when they build instruments from found objects or create them using the latest technology.

As a musical inventor, you should always keep your eyes and ears open to find new ways to turn sounds into songs. Every artist needs inspiration. That can come from nature, from the things you use in your daily life, and from listening to the work of other creative people.

You've already met some amazing musical inventors in the pages of this book. Here are a few more musicians whose ideas may point you in new and different directions. Read about their instruments, watch their videos, and catch their performances and exhibits.

- OddMusic (*www.oddmusic.com*)
- The Virtual Museum of Music Inventions (*www.musicinventions.org*)
- Experimental Musical Instruments (*windworld.com*)
- Bart Hopkin (*windworld.com/bart*)
- The Garbage-Men (*www.thegarbagemen.com*)
- Blue Man Group (*www.blueman.com*)

Also check out my Musical Inventions playlist on YouTube (*goo.gl/2ts9bO*). I'll be adding to it as new videos pop up. If you've got one you'd like me to include, contact me at my website (*craftsforlearning.com*). Good luck, and keep making music!

—Kathy Ceceri

Index

D

E

N

O

P

Q

R

S

T

U

V

W

X

About the Author

Kathy Ceceri is the author of STEAM activity books for kids and beginners of all ages, including *Edible Inventions, Paper Inventions*, and *Making Simple Robots* from Maker Media. Her other titles include *Robotics: Discover the Science and Technology of the Future* and *Video Games: Design and Code Your Own Adventure* from Nomad Press. She helped create the GeekMom blog and book, was a top contributor to Wired.com's GeekDad blog, and created more than a dozen projects for the best-selling *Geek Dad* series of books. A retired homeschooling parent, she was also the Homeschooling Expert at About.com. In addition to her writings, Kathy presents hands-on learning workshops for students and educators at schools, museums, libraries, and Makerspaces throughout the Northeast and speaks at Maker Faires around the country. Her website is *craftsforlearning.com.*